T0249864

RESOURCES FOR THE FUTURE LIBRARY COLLECTION
POLICY AND GOVERNANCE

Volume 5

Science & Resources
Prospects and Implications of Technological Advance

Full list of titles in the set
POLICY AND GOVERNANCE

Science & Resources
Prospects and Implications of Technological Advance

Henry Jarrett

Washington, DC • London

First published in 1959 by The Johns Hopkins University Press for Resources for the Future

This edition first published in 2011 by RFF Press, an imprint of Earthscan

Earthscan LLC, 1616 P Street, NW, Washington, DC 20036, USA
Earthscan Ltd, Dunstan House, 14a St Cross Street, London EC1N 8XA, UK
Earthscan publishes in association with the International Institute for Environment and Development

For more information on RFF Press and Earthscan publications, see www. rffpress.org and www.earthscan.co.uk or write to earthinfo@earthscan.co.uk

ISBN: 978-1-61726-066-7 (Volume 5)
ISBN: 978-1-61726-007-0 (Policy and Governance set)
ISBN: 978-1-61726-000-1 (Resources for the Future Library Collection)

A catalogue record for this book is available from the British Library

Publisher's note

The publisher has made every effort to ensure the quality of this reprint, but points out that some imperfections in the original copies may be apparent.

At Earthscan we strive to minimize our environmental impacts and carbon footprint through reducing waste, recycling and offsetting our CO_2 emissions, including those created through publication of this book. For more details of our environmental policy, see www.earthscan.co.uk.

SCIENCE AND RESOURCES

Prospects and Implications of Technological Advance

SCIENCE *and*

Prospects and Implications

	George W. Beadle	Henry A. Wallace
ESSAYS BY		Oris V. Wells
	Horace R. Byers	Clinton P. Anderson
		Edward A. Ackerman
	John A. S. Adams	James Boyd
		Paul W. McGann
	Earl P. Stevenson	Frederick T. Moore
		Richard L. Meier
	Willard F. Libby	Philip Mullenbach
		E. Blythe Stason
	Lee A. DuBridge	Alan L. Dean
		Philip C. Jessup

RESOURCES

of Technological Advance

EDITED BY *Henry Jarrett*

PUBLISHED FOR Resources for the Future, Inc.

BY The Johns Hopkins Press, BALTIMORE

RESOURCES FOR THE FUTURE, INC., Washington, D.C.

Resources for the Future is a nonprofit corporation for research and education to advance development, conservation, and use of natural resources. It was established in 1952 with the co-operation of The Ford Foundation and its activities have been financed by grants from that Foundation. The main research areas of the resident staff are Water Resources, Energy and Mineral Resources, Land Use and Management, Regional Studies, and Resources and National Growth. One of the major aims of Resources for the Future is to make available the results of research by its staff members and consultants. Unless otherwise stated, interpretations and conclusions in RFF publications are those of the authors. The responsibility of the organization as a whole is for selection of significant questions for study and for competence and freedom of inquiry.

This book is one of Resources for the Future's general publications. It is based on the papers prepared by authorities who participated in the 1959 RFF Forum.

Contents

v

Editor's Introduction

RECENT ADVANCES in science and technology already are strongly influencing the production and use of natural resources, and will have even larger effects in the future. Nearly everyone recognizes that this is so, but the exact nature of the new discoveries, the forces they are letting loose, and the directions in which they are leading, are not nearly so widely known or well understood. This book explores some of these difficult and critical questions.

There is nothing new in the close tie between technology and natural resources. People have been devising better ways of using resources ever since they climbed down from the trees; some of the greatest achievements, in fact, belong to prehistory —the wheel, the beginnings of organized agriculture and metallurgy, and other epoch-making discoveries. During the long interval the range of usable resources has been constantly broadened, though many particular methods and materials have been supplanted along the way. Neither is there anything unfamiliar about the way that resource technology is supported and propelled by modern science; that relationship in its modern form dates from the Seventeenth Century. The new elements are, rather, the continuing acceleration of the technological drive and the fundamental nature of some of the things science is discovering.

The world has not yet nearly assimilated the great technological gains of the Nineteenth Century whose many accomplishments included the widespread harnessing of steam power, mastery of electricity, and development of the internal combustion engine. Before the digestive process is finished there will be many further changes in farm and mine production, factory output, wage levels and work weeks, trade patterns, road-building, school systems, taxes, and almost everything else. Meanwhile new technological yeast is constantly being added to the ferment.

Consider just a few of the events of the past twenty years. Nuclear energy has been unlocked, changing uranium from a little-used resource into a prized raw material, promising to ease the strain on coal, oil, and gas as sources of energy, posing the riddle of disposal of radioactive wastes, and offering many other good or evil possibilities completely aside from its frightful military potential. New plastics and other products of what Earl Stevenson later in this book calls "molecular engineering" are challenging many uses of familiar materials like steel, copper, lead, wool, cotton, and wood, often enabling commonplace and plentiful raw materials to do work hitherto performed by relatively scarce ones. New knowledge of how clouds behave already appears to have given man some power to influence the pattern of rainfall in mountain areas; and although much work remains to be done a much more effective degree of weather control seems entirely possible. Continuing improvements in plant and animal genetics and other branches of agricultural science are bringing large increases in yield— so large in fact that for the present at least the United States is much embarrassed by farm surpluses. In the long run, however, the influence of modern genetics upon farm production may be small as compared with its influence upon man himself as the maker and user of resources. Then there are the beginnings of the exploration of outer space. Although, as Lee DuBridge points out, the direct effect upon resources during the next few years is likely to be consumption of vast amounts of raw materials in the effort to escape the pull of gravity, greater knowledge and control of the terrestrial environment already are in sight even from the first tentative probes; moreover, even after a taste of the great adventure, man's concept of his future and material resources never again can be completely earthbound.

With such continual changes in the patterns and background of resources supplies and use, there is every reason for reviewing periodically, and more frequently than in the past, developments on the scattered fronts of science and technology and to appraise their meaning. The need is sharpened by another circumstance: the sudden and wide acceptance of forward planning, with its heavy dependence upon projections. Man's interest in his long destiny in the next world is an old preoccupation; individuals have planned ahead for immediate families, and rulers for dynasties; but the concern of whole nations over what things will be like 25 or 100 years hence is a truly new phenomenon. And this is not just a passive concern. Even in the United States and other countries that make the deepest bows to both democracy and *laissez faire,* people act in the present to avert calamities or induce benefits in a fairly distant future. Many public and private decisions are being made at least partly on the basis of projections and analysis of their meaning. The value of a projection rests largely on its underlying assumptions, and in dealing with natural resources one of the key assumptions is that of the future state of technology. There are infinite possibilities for error here, from too much imagination as well as from too little. At first glance it might seem that the safest course would be to assume no change from present technology, but with today's rate of technological gain this is just about the surest way of going wrong. From what was known in 1900, with only a few uncertain horseless carriages on the bad roads, the probable 1950 demand for buggy whips would have looked quite respectable.

These reasons led to the 1959 Resources for the Future Forum, held in Washington during January, February, and March. The papers in this book were first presented there. As planned by Reuben G. Gustavson and Joseph L. Fisher, executive director and associate director of Resources for the Future, each of six sets of public lectures dealt with one area in which advances in the natural sciences are of large significance to resources and their management. Each section of this book is based on one Forum program. In the first essay of a section a leading natural scientist describes in nontechnical terms the present status of important lines of research in his area of special interest and the prospects, insofar as they can be foreseen, of where the next gains will come. In the other two essays authorities in fields other than natural science discuss the implications of scientific progress to the resources picture, each from his own viewpoint—businessman, government administrator, economist, political scientist, etc.

With the wide range of possible subjects, it was hard to confine the series to six. Two of the several interesting areas of research that had to be left out were solar energy and marine resources, including desalting of sea water. It may well be that some of the dramatic research breakthroughs of the next few years will come in one or more of these passed-over fields. All that can be said is that at the time of selection, the lines of research chosen were those that seemed among the most significant and most in need of public presentation and discussion.

Four of the topics finally chosen—minerals exploration, weather modification, chemical technology, and nuclear energy —are concerned directly with particular resources or groups of resources. The other two—genetics and outer space, which comprise the first and last sections of the book—deal more with the general climate in which resources of many kinds are produced and used.

In many ways the design of this book complements that of *Perspectives on Conservation,* which resulted from the 1958 RFF Forum. The earlier volume looked at the nation's resources situation from the standpoint of the first 50 years of the conservation movement, 1908-1958. The treatment was largely historical and from the inside, with emphasis on economics and politics. The present volume largely looks ahead, from the standpoint of outside influences upon resources from the laboratory and industrial plant. Together, the two books supply much of the broad background of contemporary thought and affairs without which no consideration of natural resources problems and issues can go very deep.

As for any generalization of what the experts think about the shape of things to come, the editor can do no more than invite the reader to read on and see for himself. The essays that follow resist pigeon-holing, for each distills in a few pages the accumulated thought and experience of men of varied viewpoints and strong convictions. Although the pattern of the book is neat and simple, the execution is not neat at all. The idea of having the scientists present the research background was generally adhered to, but their own interpretations and conjectures and those of the nonscientific commentators ranged far and wide and sometimes, to put it mildly, fell short of complete agreement.

Neither the authors as a group nor Resources for the Future, as sponsor of the Forum and this book, attempted to resolve the variety of facts and opinions or to set down any formal conclusions. Nevertheless I believe it would be difficult for anyone to come away from the whole collection of essays without some impressions of the pace and direction of advances in science and technology and of their impact on people and resources. By way of example, here are a few of the things the book set me to thinking about . . . .

On the advancing frontiers of knowledge: It seems clear that some of the current lines of research are truly seminal. Old limits and relationships are dissolving. In the future the top limit of fresh water supply in any given area may no longer be imposed by nature's hydrologic cycle of evaporation, rainfall, and runoff. New materials tailor made by chemical technology may change the whole concept of total supplies; the ultimate resource base would be measured in molecules rather than tons of ore, board feet of timber or acres in cotton. Deliberate modifications of the size and perhaps characteristics of future populations may influence future patterns of demand. Joseph Lerner, until recently of the RFF staff, once announced with a straight face that a good way to solve problems of resources shortage might be to breed smaller people; however hard he may have been pulling his colleagues' legs, the possibilities of altering human requirements are nonetheless plain. The possibilities of penetrating outer space add a new dimension to the man-resources relationship; even if the prospects are remote and uncertain they cannot be entirely discounted in a civilization that centuries ago had the imagination and audacity to fix the position of a ship at sea by taking sights on the stars.

On scarcity or abundance: The book seems to leave this classic resources question as open as it was before. Even though most of the developments explicitly discussed will work in the direction of increasing resource supplies, through better methods of discovery, extraction, and processing, there is little to suggest that technology has, or will have, the complete answer. Two significant undercurrents that don't encourage complacency run through the essays. One is the continuing pressure of rising world population and of levels of living, especially in what are now the poorer areas. Although the curve can't rise forever* and needs and tastes may change considerably over the long run, the outlook for many years to come is one of more people wanting more resource products. The second is the tendency of many technological improvements to increase the draft on resources. This is the same long-standing trend that in 1952 caused the Paley Commission to remark with some awe that *domestic* consumption of most fuels and minerals since the start of the first World War exceeded the total world consumption in all the preceding centuries.

---

* Recently Harold Barnett of RFF, simply to show where mathematical projections can lead, cited a calculation that if the world's population should increase at the rate assumed by Malthus the total weight of population in the year 3000 would exceed that of the earth. An alert reader, remembering the law of the conservation of matter, promptly asked: "What will they be made of?"

On human responsibility: The new possibilities of surmounting some of the old limits and barriers on resource supplies and use put a new burden on the modern citizen. Until recently man accepted the main characteristics of his environment pretty much as he found them. He survived, and sometimes prospered, by ingenuity, adaptability, and fortitude. These virtues doubtless will remain indispensable, but new possibilities of choice are now being added. As Edward A. Ackerman puts it in his essay, "Do we know what kind of weather or climate we should like to have if we could change it to order?" It can be argued, of course, that the difference is only in degree, since science still consists in understanding natural phenomena. But this really would be to beg the question, considering the basic quality of some of the things man is beginning to understand, the scale on which he seeks to alter his environment, and the possible side effects of his interference with processes and relationships that used to be thought of as unchanging. There is need of wisdom in making the choices as they open up. Sometimes the risks may be greater than the possible gains. George Beadle, for example, points out that while modern man already *knows* enough he is not yet by any means *wise* enough to take a hand in consciously shaping his own genetic future. More and more, as technology advances, must planners, administrators, and ordinary voters be aware of the physical and biological possibilities and limitations of their plans and aspirations; and scientists and technologists must recognize the social and economic meaning of the applications of their research.

Henry Jarrett, *Editor*
Resources for the Future

May, 1959

i

GENETICS

George W. Beadle:
MOLECULES, VIRUSES, AND HEREDITY

Henry A. Wallace:
GENETIC DIFFERENTIALS AND MAN'S FUTURE

Oris V. Wells:
AGRICULTURE'S NEW MULTIPLIERS

GEORGE W. BEADLE

Molecules, Viruses, and Heredity

IT IS AN unending source of wonderment that out of minute spheres of jelly-like protoplasm little larger than the point of a dull pin there should develop living beings like you and me— beings built of uncountable billions of molecules intricately organized and interrelated; capable of growth, adaption, memory, rational thought and communication; able to create and appreciate art, music, literature, religion, science and technology; and, above all, designed to hand down to the next

GEORGE WELLS BEADLE, professor of biology and chairman of the division of biology at the California Institute of Technology, is a Nobel Prize winner who shared the 1958 award in medicine for work in the field of genetics. He has occupied his present chair since 1946. Previously he was professor of biology (in genetics) at Leland Stanford Jr. University, 1937-46; and assistant professor of genetics at Harvard University, 1936-37. During the 1958-59 year he was on leave in England as George Eastman visiting professor at Oxford University. The most significant result of Mr. Beadle's work over the years, perhaps, is the cumulative one of helping to develop new ways of studying gene action. His specific research interests have been, successively: genetics and cytology of Indian corn (1926-31); genetics of Drosophila melanogaster, and chemistry and biology of eye pigment in Drosophila (1931-40); and chemical genetics of the red bread mold Neurospora (1940 to the present). He is a former president (1955) of the American Association for the Advancement of Science. He is chairman of the Scientific Advisory Council of the American Cancer Society, and of the Committee on Genetic Effects of Atomic Radiation

3

generation the biological and cultural inheritance that permits this near-miracle to be repeated again and again. All this from the tiny cell that is the fertilized egg of man.

If we could but expose the secrets that lie locked within this minute sphere, we would have achieved complete understanding of man, including the manner of his origin from subhuman ancestors and the nature of his destiny in an evolutionary future now unknown. Not in your time or mine, nor in the time of our sons and grandsons, will we succeed in doing this. But still the progress of modern science has been so great in this direction in recent years that it is now possible to redefine some of the most basic concepts of biology in terms enormously more meaningful than those used but a few years ago.

In this paper I want to tell about some of these new and exciting advances, especially those that can enlarge our understanding of the material basis of biological inheritance. I hope that I shall be able to make their significance understandable to readers who are not specialists in the technology of modern genetics.

Almost a hundred years ago Gregor Mendel gave the world convincing evidence that many of the traits by which varieties of garden peas differ from one another are inherited as though they were under the influence of discrete elements passed on from one generation to the next. The simple principles by which this takes place are now taught in all beginning biology courses and known as Mendel's laws.

His contemporaries were not prepared to accept such a simple and seemingly naïve explanation, but his successors a generation later rediscovered and confirmed his findings. They soon located his hypothetical elements, which they renamed

of the National Academy of Sciences. Honors for his work in genetics, previous to winning the Nobel Prize, include the Lasker Award (1950), the Dyer Award (1951), the Emil C. Hansen Prize (1953), and the Albert Einstein Commemorative Award in Science (1958). Mr. Beadle was born in Wahoo, Nebraska, in 1903. He received his B.S. and M.S. degrees from the University of Nebraska and his Ph.D. from Cornell University.

genes, in the microscopically visible chromosomes of the cell nucleus.

Classical genetics—which deals mainly with the mechanics of gene transmission—grew vigorously during the first third of the present century. It showed that the many kinds of genes of an organism were disposed in its chromosomes in linear order, and demonstrated that the basic principles of Mendelism were widely applicable to all higher plants and animals, including man. The effects of high-energy radiation in causing mutations in genes were discovered and put to good use in experimental genetics. At the same time, important beginnings were made in understanding the chemical nature of genetic material and the manner in which it controls developmental and functional processes in complex living systems.

Returning to the thought that all of us begin life as individuals in the form of single-celled eggs, it is obvious that both the manner in which we develop and the end results of our development depend on two groups of factors: first, those contained within the egg and, second, those that make up the environment.

Inside the outer wall of the egg is a layer of cytoplasm, in which is stored a certain amount of food material. In man, this is sufficient to see the embryo through its very early stage prior to the establishment of a functioning connection with the mother. In the frog it is enough for the developing embryo to reach the stage at which it can find its own food supply. The cytoplasm contains also an elaborate set of metabolic machinery for compartmentalizing and regulating the many chemical reactions that are vital to development.

Centrally within the egg cell is the nucleus. In it are the chromosomes, twenty-three from the mother and twenty-three from the father. These carry the Mendelian elements—the genes. Collectively the genes may be thought of as the directions for development—in the case of the human egg, a kind of recipe for a person.

In human beings, the environment is itself highly and specifically organized within the body of the mother during the early stage of development. Food materials and oxygen must be supplied, carbon dioxide and other waste materials must be disposed of, subtle regulating chemicals must be made available in proper amounts and at proper times, and a host of other factors must be favorable. In more primitive animals, frogs for example, the environment is less exacting, consisting of simple pond waters.

Obviously both the inherited directions and a proper environment are essential, even though the latter may be as simple as the spring water in which the frog embryo develops. But given the proper environment, the end result is largely determined by the directions. In effect, they say whether the final product of development will be a human being or a frog. If they say human being, they also say what kind—male or female; blue- or brown-eyed; tall or short; taster or nontaster of phenyl thiocarbamide, the substance that is bitter to some palates, neutral to others; and all the other inherited traits which make us men and by which we differ.

This hereditary information or recipe is precise at the molecular level. How much so is illustrated by the red blood cell pigment, hemoglobin. This is a protein, composed of 600 amino acid building blocks, of which there are twenty varieties, strung together end-to-end in long chains like dominoes. Their precise order determines that the protein will be hemoglobin rather than any one of almost unlimited other kinds of protein.

Human hemoglobin molecules have two identical halves, each with the same precise arrangement of amino acid residues. We believe that one unit of inheritance—a particular gene—carries the information used in arranging the component parts of the hemoglobin protein in exactly the right way: that this particular kind of gene represents the part of the total recipe that says how to build hemoglobin protein.

One line of evidence for this belief consists in the correlation observed between a particular modification in the gene and a specific change in the amino acid sequence of the hemoglobin protein. Thus persons who inherited an S-form of the gene possess hemoglobin molecules in which one amino acid in the sequence of 300 per half molecule is replaced by another—glutamic acid by valine, if you are familiar with the names of the amino acid components of proteins.

A person who inherits the S-form of this gene from both parents has essentially only the modified form of hemoglobin. These abnormal molecules have the unhappy property of arranging themselves in such a way that the lives of the blood cells containing them are much shortened. This results in a severe deficiency called sickle cell anemia, which is usually fatal before maturity if the best modern medical care, including blood transfusions, is not given.

It is estimated that there may be as many as 10,000 to 100,000 kinds of genes in the nucleus of the human egg, each responsible for some specific property—like the structure of hemoglobin—in each individual. These genes constitute a very large amount of information, an amount, it is estimated, equivalent to the contents of a good-sized home library of, say, 1,000 volumes.

The development of a person from a fertilized egg is in principle something like making an angel food cake. The genetic material is the recipe. It says what ingredients are necessary and how to put them together. The environment required for the right end result is the oven. If it is not right—if the temperature is too high or too low—the final product is something less than it should be. The analogy is crude but for some purposes useful.

How is the recipe for a man written? A few years ago no one knew. Today we think we do. The answer has evolved in the minds of men in a long and tortuous way, as is often the case in science. I shall simplify and streamline the story considerably.

First of all, man is not a favorable organism in which to look for the answer to this question. He is enormously complex, being built of thousands of kinds of molecules, many of which have not yet been characterized by chemists. He is not ideal genetically because of his long life cycle, and too few children—at least for scientific purposes—per family. Furthermore, his marriages are never designed with the thought of their ability to yield maximum genetic information.

For these reasons, other much simpler organisms have been chosen for investigation by geneticists. This is justified, for we now know that the principles of Mendelian inheritance are fundamentally the same in such diverse organisms as men, mice, protozoa, Indian corn, algae, fungi, and bacteria. This is, of course, what one might expect on the basis of Darwin's concept of evolution.

We can go further and find even simpler forms. The smaller viruses are subcellular, submicroscopic, and so simple chemically that they appear to be made up of only *two* classes of molecules. Thus they are far simpler than any cellular species. They, of course, reproduce only in living host cells of higher forms, some in animals, some in plants, others in bacteria.

The bacterial viruses that live in the common colon bacillus have been found to be especially favorable for genetic investigations. One of the commonly used species, called T2, is shown by the electron microscope—it is too small to be seen through an ordinary microscope—to be a tadpole-shaped structure with an angular polyhedral head and a cylindrical tail about as long as the head. It is known to have characteristics that are inherited in much the same way as are those of higher forms. The "genes" that differentiate these characters are arranged linearly in a chromosome-like structure, just as are the genes of higher forms. The two kinds of molecules present are proteins, built according to the same general scheme already described for hemoglobin, and deoxyribonucleic acids, abbreviated DNA. DNA molecules, like those of protein, are long chain-like structures. But unlike proteins they are made of nucleotide building blocks, not amino acids; and there are only four kinds of building blocks, not twenty as in the case of amino acid components of protein.

Free bacterial virus particles can be made to separate into their two chemical components by "osmotic shock"—a rapid shift from a high salt concentration to a low one in a solution in which the virus particles are suspended. With the electron microscope these can be identified as collapsed protein coats, with the virus tails still attached, and long thread-like DNA molecules.

A virus particle infects its host cell by first becoming attached to the bacterial coat by the tip of its tail. After a short time the collapsed coats are seen in electron micrographs still attached by their tails but remaining outside the bacterial cell. Since the bacteria become infected and shortly burst, with the release of many daughter viruses, the genetic material of the virus must have entered the bacterium.

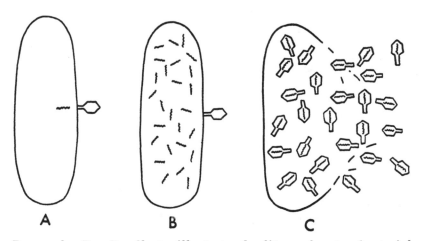

A   B   C

Drawn by Dr. Beadle to illustrate the life cycle of a bacterial virus. Genetic material of the bacterial cell is not indicated. In life the viruses are smaller relative to the host cell. A—Condition a minute or two after adsorption. The viral DNA has entered the host. B—Some ten minutes later viral DNA has replicated but protein coats have not yet formed. C—Destruction of host cell has occurred with release of daughter viruses. Under favorable conditions the virus yield is considerably higher—often 100 to 200 per cell.

Is this genetic material the DNA only, or is it perhaps DNA plus protein? This important question was neatly and elegantly answered in 1952 by Hershey and Chase of the Carnegie Institution laboratory at Cold Spring Harbor, New York. By quite simple methods it is possible to introduce radioactive atoms of sulfur into the proteins of viruses. When such labeled viruses are allowed to infect bacteria, it is seen that most of the protein remains outside with the coats. Therefore, most of the protein does not enter. If, on the other hand, DNA is labeled with radioactive phosphorus, the radioactivity enters the bacterium and very little remains outside with the virus coats. The conclusion is clear: The DNA core of the virus and very little else enters the host cell—amazingly, through the virus tail, which serves as a kind of submicroscopic hypodermic needle. The experiment can be refined even further by tracing the labeled atoms to the next generation of viruses. This shows that the DNA of the infecting viruses appears in daughter viruses but that protein does not.

Therefore it is clear that the primary genetic material of a virus particle is DNA.

Some viruses contain a form of nucleic acid called ribonucleic acid, or RNA. In tobacco mosaic viruses it has been demonstrated (in 1956) by Fraenkel-Conrat and Williams in the United States and (at almost the same time) by Gierer and Schramm in Germany that infection can be accomplished by RNA alone, the protein coats having been stripped off and discarded.

The conclusion is inescapable for viruses: The material substance that bridges the gap between successive virus generations is DNA in bacterial viruses and RNA in tobacco mosaic virus. Therefore, these molecules must somehow carry all essential primary genetic information.

How do the nucleic acids carry information and how is this information replicated 200-fold in twenty minutes, as it must be in the case of bacterial viruses? The answer is not yet clear for RNA but is much more nearly so for DNA.

However, before explaining what is known about the structure of DNA and how this is related to its function as genetic material, let us first look at the situation in higher forms—

bacteria, fungi, flowering plants, and multicellular animals. Is the primary genetic material DNA, RNA, or perhaps some other substance? This is an important question, for the evolutionary relationship of viruses to higher forms is by no means completely understood.

For bacteria the answers seem clear. In pneumococci—bacteria causing pneumonia—it is known that genetic traits can be transferred from one strain to another by way of pure DNA. Under proper conditions, this material can be extracted from one strain, purified chemically, and then incorporated into a recipient strain with a permanent genetic transformation to the genetic type of the donor.

In higher plants and animals the argument is not as direct. Chromosomes in these forms seem always to contain DNA and protein. Only during certain stages do they contain RNA; RNA thus seems excluded. By analogy with viruses and bacteria DNA is the favored candidate.

In one respect it is fortunate that DNA rather than RNA is the more general form of primary genetic material, since much more is known about its structure.

In 1953, Watson and Crick (a biologist and a physical chemist working at Cambridge University) proposed a structure for DNA that created great excitement among biologists because it seems so plausibly to account for four essential properties of genes—specificity (or information content), replication, mutation, and function. Today, after more than five years, there is widespread agreement that the structure they proposed is correct. Their work therefore represents a most significant advance in biology, perhaps the most important so far made in this century, for it bears directly on the nature of living things at a basic level. In addition, for the first time in the history of biology, the Watson-Crick structure has enabled geneticists and chemists to discuss the unique properties of living systems—replication and mutation—in concrete chemical terms and in the common language of molecular structure.

How is this remarkable molecule constructed? Although I cannot now describe it in detail, I can, I believe, indicate its most significant features in a simple manner. The unique feature of DNA as represented in the Watson-Crick structure is its double complementary nature. Two polynucleotide chains—that is, long specific sequences of the four nucleotides— run parallel but with opposite polarities, as determined by the directions in which the nucleotides are oriented in the two chains. The parallel chains are complementary in the sense that for each of the four nucleotides in one chain there is a particular but different nucleotide at the same level in the paired chain. Thus in the double chain there are four kinds of pairs of nucleotides, arranged sequentially. The members of each pair are hydrogen-bonded together in a specific way, such that only the four pairs shown below fulfill proper specifications.

The structure can be visualized in two dimensions in the following scheme:

$$\text{-A -T-C-G-}$$
$$\text{-T̈-Ä-G̈-C̈-}$$

—in which A,T,C, and G designate the four nucleotides characterized by the nitrogenous bases adenine, thymine, cytosine, and guanine.

Only a four-unit segment is represented. A single double chain may be made up of as many as 80,000 such nucleotide pairs. In their native state, the paired chains are wound into a double helix just large enough in diameter to be resolved by a modern electron microscope.

Genetic specificity is believed to reside in the sequence of nucleotide pairs. If all sequences are possible, it is obvious that the number of possible specific DNA's is 4^n, where n is the number of nucleotide pairs. In this case there would be possible 256 different four-pair segments if read in only one direction. If the Watson-Crick formulation is correct, it is proper to say that genetic information is written in the form of a four-symbol code, where each of the four possible base pairs is one of the symbols.

A basic property of genetic material is its ability to replicate in a very precise way. Thus in man it is believed that at cell division the genetic material is exactly replicated in such a way that every cell of the body receives in its nucleus the total information originally present in the fertilized egg.

With the formation of eggs and sperm a special process occurs by which for each bit of genetic information (or gene), one representative is transmitted to the next generation. For each gene it is a matter of chance whether a descendant of the mother's or the father's contribution to a reproducing individual ends up in a given egg or sperm.

We do not know in detail how DNA is replicated, but we do know from isotope labeling studies that what is observed is consistent with the following simple scheme:

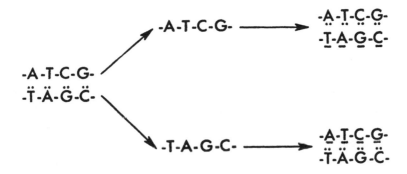

—according to which the two parallel strands separate, with the resulting half-molecules serving as templates on which are collected the proper nucleotides to reconstruct on each, a complementary half. The individual building blocks used in the process are synthesized in the cell in which the replication takes place.

Professor Kornberg and his coworkers at Washington University have succeeded in devising experimental conditions in a test tube sufficiently similar to those in the living cell so that some synthesis of DNA occurs. Since all four nucleotides are necessary and since a "primer" piece of DNA originally synthesized in a living cell must be added, it is possible that the observed synthesis represents true biological replication of the specific DNA added. If so, we shall soon understand a great deal more about this process, which is the molecular basis of all biological reproduction.

If DNA carries information in the form of a four-symbol code, it is tempting to believe that gene mutations represent mistakes like typographical errors in a message printed in the twenty-six letters of the alphabet. Anyone who has ever used a typewriter can easily see that there might be several kinds of errors: omission, addition, transposition, and substitution of symbols. It is only a hypothesis that the analogous errors in DNA replication are the basis of biological mutation, for as yet no one has succeeded in determining the sequence of nucleotide pairs in even a small segment of DNA. But it is a plausible and useful working hypothesis for which there is some circumstantial evidence and for which many chemists and biologists are now seeking direct experimental evidence.

This interpretation of mutation is consistent with the well-known fact that mutant genes replicate their kind as faithfully as did the normal genes from which they came. The mechanism described above by which DNA is thought to be replicated is independent of the sequence in which nucleotide pairs occur. Hence, useless genetic information is multiplied in exactly the same way and with the same precision as is useful information. It is as if a typist copied the recipe for our angel food cake in a purely mechanical way, reproducing any typographical errors in the original, then made a second batch of copies from her *first* copy, and so on, with never a proofreader to catch the accumulating errors. Suppose she makes five mistakes each time. Before too long the whole page would be nonsense.

Parenthetically I might say that this is the basis of geneticists' concern about increasing mutation rates in man through exposure to nuclear radiation or other man-made mutagenic agents in addition to the background radiation that always has been present. The mutant genes so produced perpetuate themselves. Since they are mostly deleterious, they are eventually eliminated by natural selection—or, more dramatically stated, by genetic death. This is the process, to return to our typewriting analogy, by which the sheets full of errors are finally thrown into the wastebasket. As yet, scientists know little about how natural selection actually works, but it does seem certain that the selective forces change over the years and that, at least in man, the process is very slow.*

*editor's note: *In the discussion period following presentation of the papers, Mr. Beadle was asked if he could clarify apparent divergencies of scientific opinion on the effects upon human populations of fall-out from nuclear explosions. He replied as follows:*

This is a large and exceedingly complex question and I can hope to do no more than summarize a few of its genetic aspects. If the products of nuclear devices—weapons, reactors, and others—are not contained, they pollute man's environment and will surely increase his mutation rate. New mutations arising in this way may be transmitted to future generations if they occur in the germ line prior to reproduction. With present methods and current levels of testing nuclear weapons, it is calculated that the mutation rate in man may be increased by as much as one per cent. This means that for every 100 mutations without testing, something like 101 may occur with it. Percentagewise, this is a small increase, but in terms of absolute numbers of new transmitted mutations per generation on a world-wide basis, it may affect as many as 200,000 individuals. Unfortunately, there is a considerable uncertainty in these calculations. Furthermore, geneticists cannot now say how these mutations will be distributed with respect to the various categories of final effects. Some will be eliminated in ways that will cause little damage in terms of human suffering or social burden. Others will surely be more serious.

The relative importance attached to the genetic hazards resulting from nuclear weapons testing tends to vary among individuals—including geneticists and other scientists—according to their attitudes toward military, political, moral, and other aspects of the nuclear weapons problem. Those who believe that ultimate survival is likely to depend on great capability in nuclear weapons, tend to minimize the genetic cost in testing them. On the other hand, those who hold that abolishment of nuclear weapons is both desirable and attainable as a means of averting global catastrophe, are inclined to use genetic hazards of testing as one of their important arguments.

That, it seems to me, summarizes the genetic side of the problem. Speak-

How is the information contained in DNA molecules used in the control of developmental and functional processes? This is a biological question of paramount importance on which many investigators are now at work. It can be restated in terms of the more specific question: How is the sequence of amino acids in a protein of hemoglobin, for instance, determined by the segment of DNA that is a gene?

A widely held hypothesis visualizes the process in the following way: A DNA molecule represents a kind of code for which there is a counterpart in the amino acid sequence of a protein molecule. This permits the transfer of information from the four-symbol system of DNA to the twenty-symbol system of protein in much the same way as the two-symbol Morse code is translatable into the twenty-six-symbol code that is our written language.

Since there are only four symbols in DNA, various groups of these must correspond to the twenty symbols in protein, again just as sequences of two, three, or four dots and dashes in Morse code represent letters. The simplest coding system that seems capable of satisfying the requirements is one in which successive triplets of nucleotide pairs in DNA correspond to specific amino acids. It turns out that there are just twenty nonoverlapping triplets possible within four symbols—provided one always reads the code from the same end of the sequence.

Thus in the sequence BDBACBADA, where A,B,C, and D now represent the four possible nucleotide pairs of DNA, the triplets BDB, ACB, and ADA correspond to three specific and different amino acids, but the triplets overlapping these, namely, DBA, BAC, CBA, and BAD do not correspond to any of the twenty protein building blocks.

ing now as a citizen rather than as a geneticist, I can say that I personally look upon the continued build-up of nuclear weapons as a most frightening business, partly because testing them creates hazards, but mainly because having them by more and more nations constitutes a risk of all-out war that seems intolerable to me. I sincerely hope, therefore, that an effective way to reverse present trends will be found—and soon.

Suppose in the hemoglobin molecule of man the middle three of the above sequence of nine nucleotide pairs were to stand for the building block glutamic acid in a particular place in the molecule, with the first and third triplets corresponding to its neighbors in the chain. If the ACB triplet were to be replaced through mutational error by ADB and this were to stand for the amino acid valine, the resulting hemoglobin would be modified in the manner observed in S-hemoglobin as compared with its normal counterpart. That is, valine would replace glutamic acid at two specific places in the molecule.

Of course in this system of coding, a segment of DNA 900 nucleotide pairs long would be the minimal section capable of specifying the complete amino acid sequence in the 300 units that are the half-molecule—three nucleotide pairs for each amino acid in the chain.

This coding system may well be too simple, but it illustrates the principle that almost surely must be involved.

It seems conceivable that the DNA-protein translation may be direct in the case of viruses, the DNA serving as a physical template or jig along which amino acids collect in a sequence corresponding to the order of nucleotide pairs.

In cellular forms, however, there is a rapidly growing body of evidence suggesting that the transfer is indirect. The DNA of the nucleus—the primary genetic information—is believed to direct the synthesis of RNA with corresponding specificity; that is, with a nucleotide sequence corresponding on a one-to-one basis with the sequence of nucleotide pairs in the master DNA. RNA differs from DNA in the structure of its four nucleotides, but each of the four of RNA is uniquely related to one in DNA.

The RNA, a kind of interpreter, moves from the nucleus into the cytoplasm where it is incorporated into small structures called microsomes. In these an RNA molecule serves as a template against which specific protein molecules are built in succession, each being peeled off the template as soon as it is completed.

Enzymes are proteins or protein-containing catalysts responsible for accelerating vital reactions. In general, enzymes are highly specific for particular chemical reactions or classes of reactions. This specificity resides in protein structure, presumably amino acid sequence. It is found experimentally that many enzymes appear to have their specific properties determined by genes, presumably through deriving their amino acid sequences from appropriate sequences of nucleotide pairs in their corresponding genes. In case of some or perhaps most genes, it appears that this may be their sole function—aside from that of serving as templates for their own replication.

If, for a gene that normally functions in this way, there is substituted a form of the gene modified by mutation in such a way that its protein counterpart lacks enzymatic activity, a corresponding chemical reaction will fail.

A special type of feeble-mindedness in man is explainable in this way. If the defective gene involved is inherited, through both the egg and the sperm, the resulting individual is incapable of converting the amino acid phenylalanine, derived from proteins in the diet, to the closely related amino acid tyrosine. Because of this block in the normal reaction, presumably because an enzymatically inactive protein is specified by the defective gene, phenylalanine accumulates in abnormally large amounts. Some of it is excreted as such. A portion is converted to a closely related keto acid known as phenylpyruvic acid and it too is excreted in large part. Nevertheless the abnormal unexcreted accumulations, it is believed, are enough to prevent normal development of the brain.

A dozen or so inherited metabolic disorders in man are explained in an analogous way. Many years ago they were called "inborn errors of metabolism" by the English biochemist and physician Sir Archibald Garrod. In some instances cures are possible. Galactosemia in man—an inherited inability to use galactose, which is a component of milk sugar—is a serious and often fatal disease in infants. If it is diagnosed early enough, and galactose-free synthetic milk substituted in the infant's diet, recovery is rapid and complete. It should be pointed out, however, that while the difficulty may be completely circumvented, the defective genes are not repaired. In the long run, cures of this kind tend to increase the frequency of defective genes normally held in check by natural selection. The increase will be extremely slow in most cases and can be humanely prevented if cured individuals refrain from reproduction.

Inherited errors of metabolism have been produced in great variety in experimental organisms like bacteria, molds, and fruit flies. Their study contributes in important ways to better understanding of gene action and the details of metabolism. They sometimes are helpful, too, in studies of the corresponding inborn errors in man and may suggest ways in which these can be alleviated by simple procedures.

Our new knowledge of gene structure promises to fill an important gap in our understanding of evolution. We can now define life in objective terms—ability to replicate in the manner of DNA, and to evolve through mutation and natural selection. Biochemists may, before long, be able to duplicate in test tubes the conditions under which "living" molecules arose on earth a few thousand million years ago. Understanding of the nature of life is thus replacing mystery.*

We now see evolution as a continuous process by which elements evolved from hydrogen; inorganic molecules and "organic" molecules arose; these interacted to produce replicating systems like DNA; virus-like systems evolved into cellular forms; these in turn evolved into multicellular plants and animals; and finally man arose, with his capacity for adding cumulative cultural inheritance to the mechanical biological

*EDITOR'S NOTE: *Asked whether new evidence of the nature of life would increase the difficulties of reconciling the teachings of science with those of religion, Mr. Beadle said:* No, I do not think our increased knowledge of the nature of living systems and the manner in which they evolved in the first place change the interrelations of science and religion in any fundamental way. We have only shifted the question of ultimate creation back in time. I might add that religion is part of human culture; it evolves, too. But creation remains a mystery as far as science is concerned.

inheritance of his remote ancestors. All this is now believed to be an orderly process in which the individual steps are not unlike those we observe today in experimental organisms. Separate mutational steps occur inevitably and those that confer selective advantages to the individuals that carry them are multiplied preferentially. Organic evolution, though unpredictable in direction, is bound to occur when the conditions are favorable.

Man's spectacular success in conquering his environment, in increasing his food supply, and in multiplying his numbers is itself evidence of the selective advantage of intelligence and the transmission of accumulated cultural knowledge. We now direct the evolution of the plants and animals that serve our many needs. Our ability to do these things ever more successfully and efficiently rises steadily. We could in fact apply the knowledge we have attained in directing our own evolutionary futures, but unless we were to do this with a great deal more wisdom than we have so far demonstrated, we would surely fail miserably if we tried.

With present rates of population growth, something will have to happen before too many generations. If we do nothing —which, even though negative, is a decision—populations will surely go on increasing at rates that will still further tax the civilizations that feed, clothe, and house them. In addition to a general increase in numbers, there will inevitably be changes in population composition, for at no time in the foreseeable future will different segments of the human population reproduce at equal rates. If the A's outbreed the B's and C's and D's, will it be because the A's are better fitted to direct man's future biological and cultural development? Or will it be because the A's are merely less responsible in controlling their numbers? The long-term consequences will be very different in the two cases.

Questions of these kinds are at once vitally important and enormously difficult to answer. They go far beyond science in their implications. Biology, including its applications in agriculture and medicine, will help in important ways to find the answers. Geneticists will have a great deal to say about such things as the necessity of genetic diversity in populations, the biological consequences of interpopulation mixing, and the effects of increased mutation rates that will result if exposure to artificial radioactivity is significantly increased over its present levels. But what is to be done, once answers to all these questions begin to be available, will have to be decided by society as a whole, and on a world-wide basis. Such decisions will inevitably require important modifications of present ways of thinking about social problems. For example, can we go on indefinitely defending as a fundamental freedom the right of individuals to determine how many children they will bear, without regard to the biological or cultural consequences?

I leave this last question for you to ponder, for I can do no more than raise it.

HENRY A. WALLACE

Genetic Differentials and Man's Future

NO AUDIENCE of general readers has ever been given such a rapid and authoritative view of both the panoramic and the detailed biochemical-genetic aspects of life as Dr. Beadle has given in his paper. In it each of us has been made to feel in his own being two billion years of the change, selection, and evolution of life all the way from virus to man.

Dr. Beadle richly deserved the Nobel Prize at least ten years

---

HENRY AGARD WALLACE, editor and plant breeder, was associate editor of *Wallaces' Farmer*, 1910-24, editor, 1924-29; and editor *Wallaces' Farmer and Iowa Homestead* (merged), 1929-33. He was Secretary of Agriculture, cabinet of President Roosevelt, 1933-40; Vice President of the United States, 1941-45; and Secretary of Commerce, 1945-46. He was formerly editor of *The New Republic*. Mr. Wallace was candidate for President in 1948. Among some of his many books are: *Corn and Corn Growing*, 1923; *America Must Choose*, 1934; *The Century of the Common Man*, 1943; *Sixty Million Jobs*, 1945; *Corn and Its Early Fathers* (with Dr. Brown), 1956. Mr. Wallace was born in Adair County, Iowa, in 1888. He received his B.S. degree from Iowa State College in 1910. Mr. Wallace while a high school boy planted corn on an ear-to-row basis and pulled out the tassels of every other row, saving only the ears from those rows to sell for seed. In 1926 he founded the first company devoted exclusively to the production of hybrid seed corn. He started work with the inbreeding and crossbreeding of chickens in 1923. The company which Mr. Wallace founded for producing hybrid seed corn took over the job of pro-

ago. He has opened up many doors between chemistry and genetics. I feel it a rare honor to comment briefly on a few of the social, economic, political, and perhaps philosophical vistas which his paper has disclosed, especially in the concluding paragraphs.

The most astounding social, political, and economic fact of this century has been the rapid increase in population, especially in the already crowded sections of the world since 1946. Many peoples are now increasing at a rate to double in thirty years instead of in fifty to sixty years as in the recent past. The first really strong impetus toward population increase began about 7,000 years ago and for some thousands of years the acceleration was rather slow. It was the domestication of food plants, especially the grains, which made possible great population increases. When men changed from a hunting, fishing, wild-food gathering way of life to a farm-village economy based on hand-seeded grains, the door was opened for the division of labor, the building of cities, commerce, banking, politics, religion, and all the rest. Following the domestication of crop plants the next really profound impact on the rate of population growth came with the discovery of America. After that, the principle of ever-expanding acceleration began to express itself more and more, and especially since 1800.

Today the average farmer in the United States is about forty times as efficient as his great, great, great grandparents of the late eighteenth century. At the moment, agricultural efficiency is outrunning city efficiency. Each year there are 3 million new mouths in the United States, and fewer farmers on the land; nevertheless farm output still increases faster than the population of the United States. The same is also true of western Europe. The critical spots today are the Near East, North Africa, southern Asia and parts of Latin America and the

---

ducing hybrid chickens in 1935 and today sells millions of baby chicks in the United States and many foreign countries. Today Mr. Wallace, fifty miles northeast of New York City, is engaged in breeding chickens, gladioli, and strawberries.

West Indies. China has 13 million new people each year and in twenty years will be nearing the billion mark. Contrast these crowded lands with Russia and the United States, both of which during the past 160 years have occupied vast areas of land which they did not have before. Russia and the United States are the two greatest "have" nations of the world. Both can use extensive methods on large farms and can thus lay the base for a much higher standard of living than the nations with five-acre farms operated largely by hand tools.

Despite their small farms and declining standard of living the "have-not" nations are outbreeding the "have" nations. World population now increases at the rate of about 48 million a year and probably 30 million came from areas where the income per capita is less than one-tenth that of the United States and where the percentage of illiteracy is more than 50 per cent. These 30 million, born into areas where the people have seen their misery steadily increasing year after year, will, as their numbers soar, become the most powerful political force in the world. Ten years hence there will be at least one and a third billion people whose standard of living, already low, will have gone down at the same time as the standard of living in the United States, Western Europe, the Soviet Union, Canada and Australia will have gone up. The terrific tension resulting from this growing discrepancy is certain, whereas the dropping of the atom bomb is not even probable. Communist countries will try to blame capitalism for the widening gap in living levels, but in so doing will be deceiving chiefly themselves. In the long run Russia is even more threatened by the population explosion than the United States. Personally I am not worried by the possible inferior genetic quality of these rapidly breeding people but I am terribly concerned about the increasing lack of opportunity for them to demonstrate their productivity in terms of hope and joyous living.

Dr. Beadle handles this situation by saying that "something will have to happen before too many generations." He is plainly worried about differential fertility lowering the genetic quality of the human race. Yet he feels that we do not at present have the knowledge to direct the evolutionary future of the race. In the more distant future he foresees the time when our biochemical genetic knowledge will warrant taking action on a world-wide basis, and he leaves us with the question of whether the day may not come when individuals will no longer have the right to determine how many children they will bear.

It is challenging when a man of Dr. Beadle's profound knowledge of the chemistry and genetics of life steps out of his purely scientific background into the broader fields where scientists may roam only at their peril. In these fields highly charged emotion usually has more power than reason. The essence of the scientific method is to move very slowly and to know everything that is possible to know about a certain small area. The new scientific truth must stand every possible criticism by fellow scientists. On this logical, truthful approach the modern technical, productive world in farm and city is based. We all respect scientists, but we find they are not at home in politics because they are tolerant and open-minded and know that all political parties have their black, white and gray sides. Knowing this, nearly all scientists retire to their laboratories and concern themselves almost solely with scientific truth rather than with social or political implications.

The outstanding exception was in the fall of 1945 and the spring of 1946, when civilian vs. military control of atomic energy was up for consideration. I know, because as Secretary of Commerce I worked with the scientists to put over the McMahon Atomic Energy Act of 1946. The atomic physicists knew, as soon as the bomb went off at Alamogordo, the nature of the domestic and international political ramifications. Their consciences made them speak out. They are still speaking out through the *Bulletin of the Atomic Scientists* on many matters which go beyond pure science.

Dr. Beadle occupies today a position in the biochemical-genetic world somewhat analogous to Rutherford or Bohr in the world of atomic physics. The Curies, Einstein, Planck,

Bohr, et al. did not envision military, political, economic, philosophical, religious, or any other consequences as they sought the truth of the atom so devotedly. Today we know their discoveries laid a foundation for the possible ending of all life on earth. Science had unwittingly made the Sermon on the Mount mandatory as an ultimate guide to international relations. Complete chemical and biological understanding of the DNA of the chromosomes will sooner or later shake our social, political, and religious life even more profoundly.

I tend to shiver when I find a Ph.D. in biochemistry writing last fall: "We will be able to plan ahead so that our children will be what we would like them to be—physically and even mentally. It will be at that point that man will be remolding his own being. Already we can produce mutants in bacterial strains; we will soon be able to control these changes; and it is not such a big jump from bacteria to plant, to animal, or to man himself."

Dr. Beadle states that we now know enough to begin directing man's evolutionary future but he questions our wisdom to do so. From a purely scientific point of view I question whether we now have, or will have within fifty years, the knowledge which would enable a genetic dictator to avoid the most serious mistakes. Undoubtedly our knowledge of DNA, the linking of specific genes with specific enzymes, the transduction of genetic material from one bacterium to another, and the work relating material in the cytoplasm to the DNA inside the nucleus and the like, has in it explosive power akin to that which characterized the early work of the atomic physicists. No genetic bomb has yet been dropped. I trust it may never be.

Just the same I hope that my grandchildren may live to see the day when hereditary and health records would be kept of all the people of the world. The kind of records I have in mind would list the diseases, the causes of death, the intellectual attainments, the abnormalities, the blood types, etc., of each person by families. These registration books would in no sense be an Almanach de Gotha or a Social Register. Nor would they ever be used by any genetic Hitler. But they would be useful to doctors, to students of genetics, and above all to young men and women of the thoughtful type who of their own free will would choose their marriage partners so as to avoid as nearly as possible the probability of hereditary disease.

The genetic constitution of man is our most precious natural resource. Probably it is not quite so strong today as it was 7,000 years ago when man first started agriculture. Forces now working at increasing speed will tend still further to weaken the heredity as distinguished from the culture of man. No great damage has been done as yet to man's heredity as distinguished from his culture by either atomic radiation or the overbreeding of the genetically inferior. On the other hand, no step has been taken to improve the genetic quality of man. If anyone even presumes to whisper in that direction he is immediately shut up by reference to the reply that Bernard Shaw is said to have made to a suggestion in applied eugenics by Isadora Duncan. (But suppose the child should have *your* brains and *my* body.) Such wisecracks express a fundamentally sound resentment against any type of genetic planning for man.

Our inborn instinct is to resist planning of all kinds. We hate a planned economy because it denies liberty. "Planned genetics" would offend the inborn essence of man even more than a "planned economy." But when it is a question of ruining gradually though surely the quality of human life, may it not be wise to furnish all the peoples of the world, generation after generation, the background information to enable intelligent young people to make free-choice the matings which will increase the genetic wealth of our planet? Mineral wealth, oil wealth, soil wealth—all are so little compared with that wealth which on the one hand is represented by culture and on the other hand by heredity. Man is on the threshold of controlling everything but himself. Soon he must start in some acceptable way on himself also, or all the rest will be as naught.

In conclusion I wish to touch most briefly upon matters for which there is no proof and for which there never will be proof. Jesus, referring to his own people, said: "I am come that they might have life and have it more abundantly." The word "life" is here used in its ultimate dimension. Some of you may remember the story attributed to the great French mathematician, Pascal, to the effect that that which is known is like a circle pushing against the "unknown." The larger the circle of knowledge, the greater the extent of the unknown. No matter how much we may learn about DNA and the control of heredity, I say that the mystery of life will grow greater, not less, as our knowledge expands. Small knowledge promotes great confidence. Much knowledge creates profound humility as the vastness of the universe from the infinitely small to the infinitely great becomes dimly sensed.

True wonderment and reverence are the beginning of wisdom. I would not for a second deter the extraordinary work of the modern biochemical geneticists, but I trust that some room is still left for that unknown factor, that semantically confusing term, "love"—love of plants, love of animals, love of families, love of God. The scientist will never have tools with which to work in this dimension but I trust that in his efforts to push the tower of Babel to the skies he does not destroy that ultimate factor which usually has so strongly characterized the parents or grandparents of most scientists. The ancestors of most scientists believed that man was created in God's image. And so as we seek to establish that our original ancestors were viruses two billions years ago, I trust we also remember the survival value for human kind of the Golden Rule. Pure materialism working through atomic energy and biochemistry will destroy that spirit which I believe to be at the heart of all creative work. Religion as well as sociology and politics must adapt itself to the new knowledge, but in so doing must retain that fundamental faith in the spirit without which all our resources are reduced to ashes. May we find the wisdom and knowledge to reconcile our past with an ever more rapidly changing future!

ORIS V. WELLS

Agriculture's New Multipliers

AFTER READING Dr. Beadle's manuscript, I first of all remembered the terms on which I was invited to discuss it: not to take issue with the author on scientific and technical grounds but rather to concentrate on the social and economic implications of his subject. This limitation is no hardship. Dr. Beadle's presentation not only agrees with everything I know about recent genetic developments, it also goes far beyond any technical knowledge that I possess and does it in a most interesting and informative manner.

ORIS V. WELLS is an agricultural economist and food expert who has been with the United States Department of Agriculture since 1929. He worked on war food problems during World War II, was chief of the Bureau of Agricultural Economics from 1946 to 1953, and helped conduct the first World Food Survey of the Food and Agriculture Organization in 1945-46. He has been administrator of the U.S.D.A.'s Agricultural Marketing Service since 1953 and is also currently on the Board of Directors of the Commodity Credit Corporation, a liaison representative to the Food and Nutrition Board of the National Research Council, and a member of the Program Committee of the Food and Agriculture Organization of the United Nations. Mr. Wells, a past president of the American Farm Economic Association (1949), was born in Slate Springs, Mississippi, in 1903, reared in southern New Mexico, and educated at New Mexico A. and M. College, the University of Minnesota, and Harvard.

29

My second task after reading the paper was to relieve the impression that I was back in my office dealing with the everyday worries of installing and operating one of the new electronic computer systems. Such systems also operate from "instructions"—instructions stored on coiled cylindrical tapes which faithfully replicate themselves or at least control everything that can possibly come out of the calculator. Such tapes also occasionally contain errors, errors of omission, addition, substitution, and transposition which replicate or otherwise gum up the works until they are eliminated.

Once having escaped Dr. Beadle's immediate and vivid imagery, I tried to remember where and when I had seen evidence of our advancing knowledge of genetics. That is, I accepted his paper as falling in the field of *basic science* and turned my attention to what we call *applied science* (assuming that *basic science* deals with the development of more satisfying explanations in terms of the reproducible or predictable truth in which scientists are interested while *applied science* suggests more efficient or less costly methods of doing the things which consumers, businessmen, administrators, or other practical everyday kind of people want done).

I remembered the Mexican cattle which I grew up with in the Southwest. They were magnificent brutes for their day and land in which they had to survive. Descended from the strays which Francisco Vasquez de Coronado dropped on his way north from Mexico in search for the Seven Cities of Cibola in 1540, they reached a high state of perfection over the next three and a half centuries, defending themselves against such predators as wolves and mountain lions, adapting to the drought-stunted vegetation of the native ranges and avoiding, so far as possible, the increasing attention of Indians and cowboys who found them to be a superior source of a harsh, tough meat and Spanish rawhide.

And then within a space of not more than thirty-five years, they vanished. Why? Simply because the American public as a part of its rising standard of living wanted more meat of a different kind.

So southwestern ranchers began to change the genetic instructions which served as the replicative basis of their industry. What was more difficult, however, once the ranchers had embarked on the process of purposive selection, was the discovery that they also had to change the handling methods and environmental conditions to which the new beef breeds were exposed if they were to successfully compete in the nation's meat market.

I remembered thirty years ago assisting a young plant breeder to devise and run a series of correlations as a first step in analyzing the biometrical characteristics of existing seed stocks of cotton then being used in the Mesilla Valley of New Mexico. That year, 1928, the average per-acre yield of cotton lint in New Mexico was 325 pounds: last season, thanks to the outstanding plant breeding job Stroman did and the farmers' ability to nourish the new, high-yielding strains with adequate supplies of irrigation water at critical times along with the right fertilizers, the per-acre yield was 820 pounds of lint.

I remembered, living as I do in an administrative world much interested in improving management, that the best discussion of efficient management techniques which has come to my attention deals not with business or government executives, but rather is an essay on the management of hybrid chickens, assuming that the farmers' main interest is in maximum egg production per unit of feed and related expense. Once again I found myself recalling a process that had to do with a change in genetic instructions, with purposive rather than natural selection, and with the provision of the appropriate environment for the new genetic material.

What I am trying to say is that one of the great scientific breakthroughs of my time has to do with an interrelated complex, with genetics, with nutrition, and with the related biochemical sciences which deal with plant and animal growth and health. These are agriculture's new multipliers, the things which make it possible for farmers to produce more and more end-use product with less and less labor and, relatively speaking, less and less land. The process involves not only the hunt for superior genetic material; it equally involves the creation and maintenance of the kind of environment, with a high emphasis on nutrition, which will best develop the full potential of the superior material.

Consider hybrid corn. Here was an innovation which within a few years toward the end of the thirties added 20 per cent—500 million bushels—to the American corn crop without calling for any more labor or a single additional acre of land. More recently, as better adapted hybrids have been developed and farmers have paid more and more attention to the handling and feeding of their corn plants, the gain has increased. With 20 per cent fewer acres, farmers now produce an extra billion or more bushels of corn.

A second spectacular example, often referred to, has to do with broilers and their feed conversion ratio. Feed used per pound of edible product, as well as the necessary feeding time, have both been cut by more than one-third over the last twelve or fifteen years, and the end is not yet in sight.

Similar, less spectacular things are happening clear across the farm field. Whereas the traditional American way of increasing farm production has been to apply more and more labor to more and more land—with increasing emphasis since the invention of the reaper in 1831 on the substitution of machinery for direct labor—the new process produces more and more from the same amount of land usually without any increase in the amount of labor required per acre or per farm.

There is no need to quote detailed statistics of increasing farm productivity since the mid-thirties. Simply remember that a third *fewer* farm people produced well over 50 per cent *more* farm products in 1958 than in 1938. On the other side, total United States population increased a little over a third while the average per capita American diet improved quality-wise (measured in terms of what consumers wanted) by at least 10 per cent, even though our food exports increased significantly over the same twenty-year period.

How much farther can the process go? For how much longer can science and technology, along with conservation and wise use of our renewable natural resources, assure us of an adequate food supply in the United States? In the world?

These are certainly appropriate questions. Our agricultural scientists seem to think that the recent gains are not only permanent, but also that research, provided it receives the support and attention it needs, can keep food supplies in line with United States population for a long time ahead. Further, we are all aware that the foreign aid or assistance programs of our various foundations, governments and international organizations are all stressing technical assistance activities designed to introduce new and improved agricultural practices over all the underdeveloped, densely populated areas of the world.

Nevertheless, Dr. Beadle ends on an ominous note, observing that "with present rates of population growth, something will have to happen before too many generations," and asking whether we can "go on indefinitely defending as a fundamental freedom the right of individuals to determine how many children they will bear, without regard to the biological or cultural consequences."

The pertinence of this observation is emphasized by the recent revision in the Census Bureau projections of United States population and the recent United Nations study, *The Future Growth of World Population.*

The revised population projections for the United States suggest to me that our population may, barring great misfortunes of the kind that we surely intend to avoid, increase as little as one-half or at the other extreme more than double over the next forty years. I myself would lean toward the low side of this range but I also believe that modern science and our increasing concern with resource management will allow us to produce the food to give our people good diets whether we end up with 265 or with 400 million people in the year 2000.

Meanwhile, the United Nations study indicates that world population was 1.25 billion in 1860, 2.5 billion in 1950 and that it may (based on what is labeled the "medium assumption") somewhat exceed 3.75 billion by 1975 and reach 6.25 billion by the year 2000. Such estimates make it easy to understand the increasing interest in population policy, including the current advocacy of policies designed to slow down the increase in population in India and China, as well as efforts to estimate the world's carrying capacity or the extent to which technical and organizational responses to different environments are likely to permit further increases. The most disturbing aspect of the whole matter lies in the fact that the year 2000 is such a short time ahead.

Faced with this kind of a situation, I suppose the medieval scholar would have assumed that the end of the world was near at hand and then found satisfaction in the fact that it was given to him to be the most advanced of men. We have no such easy solution. So we tend to choose sides, to divide as between those who feel that human ingenuity will somehow prove equal to the tasks ahead and those who doubt.

Personally, I have a considerable faith in human ingenuity, especially if we can solve the problem of simply getting along with one another. And in any event it seems to me that the central problem is one of maintaining what we call the "open" society, the "adventurous society" that refuses to regard "ancient customs as magical or sacred, that views its institutions as man-made for human purposes, that welcomes variety and change instead of enforcing rigid conformity, and that accordingly provides its members with personal opportunities and responsibilities beyond mere obedience" (Herbert J. Muller, *The Uses of the Past,* Ch. III). So long as we (and other nations too, of course) maintain such a society, I suspect human ingenuity will prosper—that we shall have enough to eat.

ii

WEATHER MODIFICATION

Horace R. Byers:
WHAT ARE WE DOING ABOUT THE WEATHER?

Clinton P. Anderson:
TOWARD GREATER CONTROL:
HIGH RISKS, HIGH STAKES

Edward A. Ackerman:
WEATHER MODIFICATION AND PUBLIC POLICY

HORACE R. BYERS

What Are We Doing About the Weather?

ON NOVEMBER 16, 1946, Vincent J. Schaefer of the General Electric Research Laboratories, flying in a light airplane over western Massachusetts, dropped some pellets of dry ice into a moderately thick bank of clouds and a few minutes later observed streaks of snow coming out of the base of the clouds. Without belaboring the point as to whether or not this snow might just have happened to fall naturally at this particular time, and without giving credit to the rainmaking quackery or pagan ritual practiced by man since the stone age, we can set down this date as the beginning of scientific weather modification.

HORACE R. BYERS is chairman of the Department of Meteorology at the University of Chicago. Although he has made important research contributions in several areas of the atmospheric sciences, he is best known for his exhaustive studies of thunderstorms carried out in a large government interdepartmental research project which he directed from 1946 to 1950, and for his leadership since that time of a group of scientists in Chicago and in Arizona studying the microphysics of clouds and precipitation. He is a past president of the American Meteorological Society and of the Section of Meteorology of the American Geophysical Union, and at the present time is vice president of the International Association of Meteorology. He was elected to the National Academy of Sciences in 1950. Born in Seattle in 1906, Mr. Byers received his A.B. from the University of California (Berkeley) and his Sc.D. from the Massachusetts Institute of Technology.

In the years since that time progress has been distressingly slow, mainly because most of the effort went into trying to apply at once principles that were too poorly understood instead of into performing the research fundamental to an understanding of the atmospheric processes involved. However, some progress is being made and we can only view the prospects of weather modification with optimism. As we find ourselves releasing amounts of thermonuclear energy as large as those released by small tropical storms, there is reason to hope that weather modification, not just cloud modification, will come within our grasp.

In taking stock of the present state of the science, we shall be concerned mostly with cloud modification and its derivative, rainmaking. That is the area, little known though it still is, in which by far the most work has been done.

Figure 1. Cloud droplets compared in size with a portion of a raindrop, all greatly magnified.

Before Schaefer's cloud-seeding test, the efforts of a number of people over many years had resulted in a fairly good conception of the natural process of precipitation from clouds. Schaefer made use of this in his experiments. As Figure 1 shows, the concept was far from obvious. The small circles represent on a highly magnified scale typical spherical cloud water droplets having a diameter of 25 microns (2.5 x 10⁻³ cm or about 1/1,000th of an inch) while on the same scale a raindrop, diameter 2,500 microns (2.5 mm or about 0.1 of an inch), can only be partly shown at the bottom of the figure. A volume computation readily shows that it takes a few million cloud droplets to make one raindrop. The cloud droplet remains in the cloud or evaporates around the boundaries and base of the cloud because it has a terminal velocity of fall of only 1 cm (about ⅜ inch) per sec. The raindrop has a terminal velocity of fall of 7.5 m per sec or about 23 ft per sec so that, although it may be held in the cloud for a short time by updrafts approximating or exceeding its terminal velocity, it must fall out as rain.

Thousands of samples of drops, taken from clouds by aircraft and by other means, have been measured. The size distributions reveal a very curious fact: that the sizes are quite uniform and small in all clouds, except that just before precipitation starts, drops of rain size begin to accumulate very rapidly. All at once and rapidly the water equivalent to that contained in some million cloud drops goes into a single raindrop. As a rough estimate we can say that one raindrop forms in every one-third cubic foot of the active part of the cloud. In a fairly small cloud volume represented by a cylinder 3,000 feet in diameter and 3,000 feet in vertical dimension we would have 21 billion cubic feet and 63 billion raindrops formed from the water equivalent of some million times that number of small cloud droplets. This can all happen in five minutes in a cumulus cloud growing to a thundercloud or in thirty minutes or more in milder forms of rain.

What causes this transformation in a cloud? The Swedish meteorologist Tor Bergeron in 1933 outlined the natural process which Schaefer took advantage of fifteen years later. It was based on two facts that were observed and well known for many years:

(1) Water drops have the characteristic that they become *undercooled* many degrees below the standard freezing point of water, that is, they remain in liquid form at subfreezing temperatures. Clouds undercooled 10 to 20 centigrade degrees (to temperatures of about +15 to −5 Fahrenheit) are to be expected in the atmosphere and −40 Fahrenheit is possible.

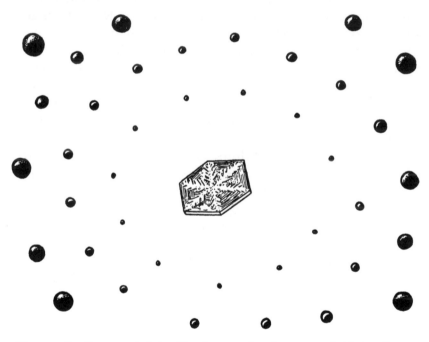

Figure 2. Ice crystal in the form of a hexagonal plate shown among undercooled droplets. Nearby droplets have evaporated partially and the vapor is being deposited on the crystal.

(2) At these subfreezing temperatures the vapor tension (the force per unit area driving the molecules to evaporate into gaseous, vapor form) is greater for undercooled water

than for ice. The difference is greatest in the temperature range −10 to −15°C (+14 to +5°F).

Bergeron reasoned that as the undercooling proceeded a few cloud particles would begin to appear in the form of ice so that there would develop a mixed cloud, consisting partly of undercooled water and partly of ice or snow crystals. The pressure driving the molecules off of the undercooled drops would be greater than that required to condense them upon the ice particles; or, to put it another way, the cloud space would be unsaturated with respect to the liquid water and supersaturated with respect to the ice. The ice crystals would feed on the vapor extracted from the drops, as in Figure 2. The ice particles would then grow at the expense of the liquid droplets until they fell as snow, melting on the way down to form rain. At the same time they would be growing by collision and coalescence with cloud droplets encountered en route to the bottom of the cloud.

Before the rates of growth by the Bergeron process could be computed quantitatively, a considerable amount of additional information about diffusion growth and collision-coalescence had to be obtained. The computations made by Henry G. Houghton of M. I. T., in 1950, showed that the rates of growth were sufficiently rapid to agree with observed rates of development of precipitation.

What Schaefer did to the cloud with his dry ice pellets in 1946 was to make a mixed cloud out of it before it was ready to become one of its own accord. Each dry ice pellet chilled a small streak of cloud air in its wake as it fell, to such a low temperature that undercooling was no longer possible. As he had already demonstrated in a clouded cold box, streaks of ice crystals developed in the wake of the pellets. He estimated that a single pellet 1 cm (approximately $\frac{3}{8}$ in) in diameter could form, under the most favorable conditions, some 10^{16} ice crystals which, if each could grow to the size of a small snow crystal, would amount to 300,000 tons of snow. Actually, the process is nowhere near that efficient and the ice crystals probably would not have access to that much water in the cloud. Nevertheless a large magnification factor is available.

Schaefer found that any cold object, such as a rod cooled in liquid nitrogen, would produce the same effect as the dry ice pellets if it was waved through an undercooled cloud in his cold box. Schaefer's basic discovery was that −40° is the lowest temperature to which water can be undercooled before it creates ice almost instantaneously. The practical inference was that anything cooler than that tossed into an undercooled cloud would guarantee a mixed cloud and the start of the Bergeron process of precipitation.

Silver-iodide seeding also was developed as a technique by the group at General Electric working under Irving Langmuir. Aware of the phenomenon of formation of crystals of one substance on other crystals of the same or *similar,* crystals, Bernard Vonnegut found from tables that the surface structure of the crystal of silver iodide was closest to that of the water (ice) crystal and therefore reasoned that it could serve as an artificial nucleating agent for ice crystals. This effect was demonstrated in the cold box.

Although it is not effective in conditions warmer than about −5°C while dry ice is good practically up to the freezing point of water, silver iodide has a great advantage over dry ice. It does not have to be dropped from an airplane. It can be carried into a cloud by turbulent diffusion, for it can be generated in the form of fine smoke particles which, in some cases, can be lifted by the air currents from ground generators into the clouds.

It was recognized by Bergeron that rain occasionally falls from clouds no part of which are colder than freezing. This was known to be true especially in the tropics. During and shortly after World War II evidence was accumulated showing that this "warm" rain initiation was more prevalent than had previously been recognized. Albert H. Woodcock of the Woods Hole Oceanographic Institution reasoned that precipitation could be initiated in this way through the action of "giant" sea-salt particles.

All cloud condensation takes place on suitable nuclei, preferably on those large enough and of such a chemical composition that little or no supersaturations (relative humidities greater than 100 per cent) are necessary in order for cloud droplets to grow on them. There appears to be a more than ample supply of such nuclei for condensation in the atmosphere at all times. There is a scarcity only of nuclei for forming ice crystals.

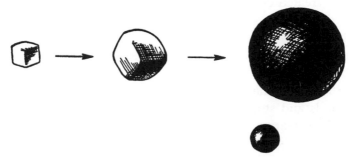

Figure 3. A "giant" nucleus composed of a substance such as sea salt (left) attracts a coating of water to it (center). As the particle reaches saturation in the cloud, it has grown to a small raindrop (right) which is large compared with the prevailing cloud droplets, one of which is shown immediately below it.

The "giant" nuclei studied by Woodcock are relatively few in number but they absorb water so fast that immediately upon reaching the saturation conditions in a cloud they have grown into drops of a radius of greater than 50 microns (about 2/1,000ths of an inch) with a terminal velocity of fall of about 30 cm per sec (1 ft per sec). (See Figure 3.) This gives them enough relative motion to collide and coalesce with other smaller droplets and thus to grow into a small raindrop.

Louis J. Battan of the University of Chicago studied summer clouds in the United States by radar and found that in the majority of them the first rain (the first radar echo) to form inside the clouds did so underneath or very near the freezing level so that the initiation of rain could not have been by the Bergeron ice-crystal process. The clouds subsequently grew to heights where ice forms would be present.

The artificial stimulation of rain in warm clouds can be accomplished by putting more giant salt nuclei into them. A more direct way, used by our group at Chicago, is to fly through them with an airplane spraying out water drops in sizes greater than those of cloud droplets and thus starting the collision-accretion process at once without requiring growth on the salt. When we treated clouds in this way in the Caribbean, we were able to double the statistical probability that rain would fall from them.

Both processes—those of mixed clouds and of warm clouds—have been described in terms of *initiation* of precipitation. Yet artificial stimulation of rain appears to increase rainfall from clouds and storms that are already producing rain. Apparently the treatment adds to the total by insuring that even the weak clouds or weak portions of active clouds become favorable for precipitation.

Irving Langmuir and the General Electric Company in 1946 realized at once that they had made a discovery of great potential importance to mankind and through Langmuir's personal enthusiasm and his company's efficient public relations office, the world knew about it very quickly. Because of the economic importance of water, here was a scientific discovery that was put to use without the customary time lag between research and practice. Throughout the world, and especially in the United States, commercial rainmaking sprang up almost overnight. Unscrupulous operators in some instances engaged in the business, making claims that were without validity. Even today this activity goes on unregulated, except by some loose requirements of certain states. The United States Weather Bureau, trying to put the brakes on runaway developments, was wrongly accused of trying to prove that rainmaking would not work. Most meteorologists were cautious about accepting the earlier so-called "proofs" and were even accused by their more optimistic colleagues, especially scientists in other fields, of opposing weather control on the grounds that it would produce technological unemployment of weather forecasters.

After more than a decade, the question of the practical efficacy of cloud seeding is only partially resolved. The area of doubt is made up of (1) doubts as to the statistical design and interpretation of tests and (2) doubts about the physics of natural and artificial precipitation.

The statistical questions apply to the direct testing of the efficacy of the chemicals or other substances put into the atmosphere to increase rain. This problem is much like the testing of a new drug or vaccine. It would not have been a good test of the Salk polio vaccine to have given it to one hundred children and then to have concluded that the vaccine prevented polio because the hundred children did not contract the disease. Similarly, we would not accept as proof of rain-making the fact that it rained on all of one hundred dark, cloudy days when silver iodide was put into the air. We should not even conclude that a marked increase of rain over a long period of cloud seeding could result from the seeding. It could as likely have happened naturally.

Herein lies the crux of the problem. The natural variability of precipitation both in time and in space is so great that artificially produced variations, if any, may not be detected in the maze of natural ones. For example, how can we recognize a man-made increase of 15 per cent in this year's rain over last year's when the natural variability from year to year is known to be 50 per cent? Or, after a certain seeding period, how are we to know what the rainfall would have been without seeding? This quantity has to be known before comparisons can be made for purposes of proof.

These questions, difficult as they may seem, can be answered with a reasonable degree of confidence by properly designed tests and rigorous analysis methods. Statistics, as a field of study, has been advanced in recent years to the point where otherwise hidden significance can be obtained from certain tests involving mixed and highly variable data and the results stated, for example, with computed probable error. The results must show figures significantly larger than the probable error; otherwise the test must be extended, redesigned, or abandoned as insensitive. Another possibility might be to devise new statistical tools which would reduce the probable error.

There is an area of disagreement among scientists concerning the statistical significance of different tests. The Advisory Committee for Weather Control, appointed by the President of the United States, wrestled for three years with the problem of evaluating the results of commercial seeding operations. Its final report, issued early in 1958, concluded that: "For the orographic [where cloud formation is influenced by mountains] and semiorographic projects—all of which were on the West Coast—the evidence indicates that cloud seeding has produced an average increase of 10 to 15 per cent in the precipitation from seeded storms . . . . For the nonorographic projects, the same procedures and testing methods did not detect any change in the precipitation that could be attributed to cloud seeding. But the fact that no increase was detected does not warrant the conclusion that there was no increase—the increase may have been too slight to measure by the statistical methods used . . . . Meanwhile, in the absence of any clear-cut statistical conclusions, a farmer in a Mid-western State is still faced with the question: Should he buy commercial cloud-seeding services?"

Since commercial operations are conducted to furnish a service rather than a test, it is obvious that more information could be gained from scientifically designed experiments.

Earlier, federal government departments and universities had conducted a "crash" program of research tests on artificial cloud nucleation in 1952-55. The results were published in 1957 in "Cloud and Weather Modification" (*American Meteorological Society, Meteorological Monographs,* Vol. 2, No. 11). Those which have a direct bearing on the present discussion may be summarized as follows:

Positive results

Dissipation of the portions of thin, stratified layers of undercooled clouds treated by dry ice or silver iodide. The snow that is made to precipitate removes the water.

Initiation of rain in small tropical cumulus clouds by spraying water into them.

Uncertain results

Increase of precipitation in western Washington by dry-ice treatment.

Modification of summer cumulus in the Mississippi Valley by dry-ice or water-spray treatment.

Undetectable effects

Large-scale energy releases in developing cyclones by dry-ice seeding.

The government in this case was mainly interested in looking for spectacular results, urged on by extravagant claims by a vociferous and strong group of scientists. The program was reduced to insignificance when two years of field work showed "minor" effects.

It should be pointed out, however, that what may look like an "insignificant" increase in precipitation can be of great economic importance. A 10 or even a 5 per cent increase in rain or snowfall in certain areas at certain times of the year can be worth millions of dollars to groups of water users.

Measurements show that there are occasions when there is a shortage of natural ice nuclei. The frequency of such occasions is largely unknown. Thus rainmaking experiments are in the position of releasing nuclei without a knowledge as to whether they are correcting a natural deficiency or are creating an oversupply which may be equally detrimental to the formation of precipitation. Some investigators have suggested that hail and lightning could be suppressed by producing an oversupply of nuclei. This area of our knowledge demands the attention of competent investigators.

Recent investigations show that the microphysics of individual particles inside a cloud represents only half of the story of natural precipitation. Meteorologists are becoming increas-

ingly aware of the importance of the meteorological, orographic, and seasonal factors in determining natural precipitation. For example, the lifetime of most cumulus-type clouds
is about equal to the period of time required for the development of a precipitation particle under natural conditions.
Thus precipitation is impossible from many clouds even
though the internal physical and chemical conditions (such as
the number of ice nuclei) are perfect for such development.
The duration of a given cloud and the nature of the air
motions in a cloud are strongly dependent upon the character
of the underlying topography. The orographic clouds which
are dependent upon mountain flow for their formation are
likely to be different inside from those which form over flat
terrain. The effect of these differences in determining the
seedability of natural clouds is not understood, although the
meager evidence available today indicates that this effect may
be very profound. Unfortunately, studies of the conditions of
the cloud environment which are important to the determination of natural precipitation have not kept pace with studies
of the microphysics and microchemistry inside a given cloud.

In order to know which natural process needs stimulation, we
need to know which of the two mechanisms—ice-crystal growth
or drop coalescence—is going to occur under given conditions.

Figure 4 shows the percentage of clouds producing radar
(rain) echoes in terms of heights and temperatures at the cloud
tops in three different geographic regions—the Caribbean, the
Middle West, and the Southwest as recorded by L. J. Battan
and R. R. Braham, Jr., in "A Study of Convective Precipitation
Based on Cloud and Radar Observations" (*Journal of Meteorology*, Vol. 13 [1956]). The temperatures are above freezing in
all the Caribbean clouds, but in the Middle West and the
Southwest most clouds reached to slightly subfreezing temperatures before rain echoes appeared. This type of "cloud
census" tells whether the ice-crystal or all-water process predominates. The geographical differences are apparent, and the
advisability of different seeding techniques in different areas
is indicated.

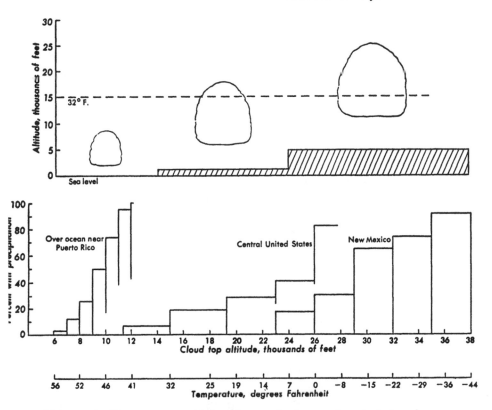

Figure 4. Percentage of clouds producing radar (rain) echoes in terms of heights and temperatures at the cloud tops in three different geographical regions.

Laboratory experiments have shown that coalescence is markedly enhanced by the presence of an electrical voltage through the cloud. Thunderstorms are known to have electric fields of thousands of volts per inch of space. Cumulus clouds have lower values, but they are apparently high enough to increase coalescence rates. Studies of these effects in the clouds are pertinent to the subject of rain inducement.

Clouds as entities—their structure and dynamics—need to be studied in order to get an idea of how the water moves through them in the form of droplets and ice crystals. We should find the true reason why one cloud produces rain while another like it in appearance does not. Temperatures and humidities outside as well as inside the clouds enter into this class of problems. Measurements of droplet sizes, liquid-water contents, and other characteristics are necessary. Nuclei of all sizes play a role which must be explored, especially the role of the giant nuclei in producing drops that can start the coalescence process.

Studies at the University of Chicago have shown that in humid areas only 10 per cent of the water that participates in an active thunderstorm reaches the ground as rain. In arid regions the percentage is less. Clouds are thus seen to have a very low precipitation efficiency, but values have not been established for clouds other than thunderstorms. We would like to know if it is possible to increase this efficiency and thereby increase precipitation at the ground.

Clouds in mountainous areas lend themselves particularly well to study and possible modification. Mountain ranges help the formation of rain over them, but it is not known exactly how. Are the clouds formed over mountains mainly by forced ascent of the air as it strikes the mountain range, or does the absorption of the sun's rays by the mountain cause rising bubbles of warm air? Research at the University of Chicago and the University of Arizona is giving an answer to this question.

Climatology, the study of conditions shown by past records, enters into the problem when we ask ourselves such questions as: What are the geographical differences in the relative importance of the ice-crystal and coalescence processes? What are the important differences between clouds over mountainous terrain and over flat areas?

Statistical problems in artificial cloud seeding have been stressed on the preceding pages. Some physical problems have been given only partial solutions. Is there sufficient time for certain artificial effects to occur in rapidly developing clouds? Is salt as effective as water spray in certain types of warm clouds? Under what conditions does ground-generated silver iodide reach the subfreezing cloud levels? What are the relative merits of airplane and ground seeding?

It has been suggested that, since hail is an extreme case of riming of an ice particle, hail could be suppressed if cloud liquid drops at subfreezing temperatures could be eliminated. If all the water at these temperatures could be nucleated to form ice crystals, there would be no droplets left to freeze to the falling kernel of a hailstone, and hence no hailstone. What would be the chances of success of such a method? Would the amount of silver iodide required be so great as to make the cost greater than that of hail insurance? Would the seeding have to be done by airplanes? What would be the design of an experiment to test this hypothesis?

From studies made in Sweden, Tor Bergeron has concluded that mountain clouds in that country are the most suitable for seeding. Australian seeders have reported greater success in seeding mountain clouds than with similar treatment over flat lands. Mountains are important because of their naturally high precipitation, their snow and ground storage, and their elevation, which permits power to be obtained from the stream outflow and which obviates the necessity of pumping.

From the preceding discussion it is clear that the "seed-ability" of clouds is variable. Geographic and time variations are noted. One of the needs in cloud physics is to develop an index of "seedability" and apply it to various cloud situations. Such an index, when combined with station records of cloud occurrences, would provide information concerning the number of "seedable" cloud situations to be expected at a place during any given time. This information could be used in determining in advance the probable economic value of a proposed seeding program. Existing weather records do not make this determination possible.

The reader is no doubt familiar with—perhaps sometimes wearied of—pleas from scientists for more support of basic research. But the need is real, and possibly nowhere is it greater than in cloud physics, a field in which people tried to run before they had learned to walk. The consequences have

been disconcerting, to say the least. Now we need desperately
to go through some of the beginning steps.

There are other possibilities of weather modification, even
more far-reaching than the inducement or inhibiting of precipitation, but even less explored. Before concluding, let me
at least mention a few of the interesting avenues that scientists
already have begun to follow up.

My colleague Harry Wexler of the Weather Bureau has
proposed an artificial way of altering the radiation balance of
the earth. His idea is to use thermonuclear explosions to make
a tremendous eruption of steam from the Arctic Ocean. Ten
explosions of ten megatons each fired from a suitable depth
would create enough steam to produce, in winter, an ice-crystal
cloud with a density of one particle per cubic centimeter
through the entire polar troposphere. In these polar regions,
north of the jet stream, the circulation is such as to keep the
cloud more or less confined there. Since in those latitudes in
winter the outgoing radiation predominates greatly over the
incoming solar radiation, the effect of the cloud would be to
act as a blanket to the outgoing radiation and prevent the
ground from reaching its normal low winter temperatures.
Wexler points out that the effective radiating surface would
then be at the top of the cloud, that is, at the tropopause. He
computes that such a cloud during the four winter months
would eliminate 80 per cent of the average radiation loss. The
amount of heat saved would be capable of warming the
atmosphere over the entire globe an average of 1.3°C (2.3°F).
Confined to a smaller part of the world, the heating would be
proportionately greater.

The main effect of carrying out Wexler's scheme would be
to accelerate the present rate of disappearance of the Arctic
ice pack. But he reckons with some undesirable side effects,
too. The Arctic cloud would make the region unsuitable for
flying activities during the four months in question. Also the
amount of snow falling on mountain glaciers would be increased, causing these glaciers to advance over inhabited or

otherwise valuable areas. The storm tracks, according to Wexler's reckoning, would be changed in such a way that there would be less winter precipitation in the belt 35 to 50 degrees north and more between 50 and 65 degrees.

Henri Dessens, of the Observatory of the Puy de Dôme in France, has created convective clouds of rain-producing potentialities by the heat from large, controlled brush fires.

A host of ideas has been brought forth for changing the climate, some beneficial, others detrimental, some impractical, others having an element of practicality. Before we start melting Arctic ice with thermonuclear energy, however, we should be sure that we know what the side effects will be.

CLINTON P. ANDERSON

Toward Greater Control:
High Risks, High Stakes

TODAY AS WE trace in the heavens the flight of our rockets and satellites, as we peer into the heart of the atom to uncover the innermost secrets of life and matter, so do we also begin to note glimmerings which may give us control over our environment. We are finding what some day may be the means of escape from the ageless and inexorable grip of climate upon the life and fortunes of this planet.

The history of man's attempts at weather control is both long and disappointing. A half-century ago, rainmaking was attempted in the high-plains country of Texas by the firing

---

CLINTON P. ANDERSON, United States Senator from New Mexico since 1948, is alternate chairman of the Joint Committee on Atomic Energy. After four years (1918 to 1922) as a reporter and editor in Albuquerque, New Mexico, he went into the insurance business, and has been the owner of a general agency in Albuquerque since January 1925. In 1932-33 he was president of Rotary International. He was Treasurer of the state of New Mexico, 1933-34; administrator of the New Mexico Relief Administration in 1935; field representative of the Federal Emergency Relief Administration, 1935-36. He was a member of the 77th to 79th United States Congresses from New Mexico at large; Secretary of Agriculture from June 1945 to May 1948. He was elected to the Senate in 1948 and was re-elected in 1954. Senator Anderson was born in Centerville, South Dakota, in 1895, and was educated at Dakota Wesleyan University and the University of Michigan.

of cannons when the storm clouds were near. More recently, truly scientific efforts have been made to learn more about the life story of precipitation, particularly rain. Horace R. Byers, in his excellent paper on weather modification, discusses the mechanics of rainmaking as we know them today. He stresses that we know relatively little about this phase of the weather process and in this he is joined by nearly every person who has studied the subject. Yet he holds forth the promise that rainmaking ultimately will be understood and utilized to the benefit of mankind.

Very soon after the conclusion of the Second World War, two scientists, Dr. Irving Langmuir, a Nobel prize winner, and his assistant, Dr. Vincent Schaefer, succeeded in creating a snow storm by dropping dry ice pellets from an aircraft into a supercooled cumulus cloud. There followed then over the years the series of precipitation experiments that is outlined and evaluated by Dr. Byers in his paper.

Within the past year, the Russians and Americans have put into orbit a series of earth satellites. Aside from their technological and scientific value and interest, we have heard speculation upon their possible usefulness as weather observation stations and television transmitters and relay points. The United States, according to published reports, is planning to use television cameras in satellites to report on cloud formation and location around the world.* Such knowledge would be of value in preparing long-range weather forecasts and these in turn could be very helpful to men everywhere.

Dr. Byers observes that immediately following the Langmuir-Schaefer discoveries a very large effort at rainmaking was undertaken by commercially-inspired organizations before there

---

*On February 17, 1959, less than a month after Senator Anderson delivered his paper, the United States launched a Vanguard satellite equipped to register and transmit data on cloud formations. *The New York Times* of February 22 reported that late in 1959 the United States planned to launch a satellite equipped with one or more miniature television cameras that could photograph the earth's cloud cover with far clearer definition than can the two photo cells of the February satellite.

was sufficient understanding of the principles involved. The shift in emphasis, he says, was detrimental to the basic need— better understanding of the rainmaking process, its applications and the implications of each. He finds progress in these directions distressingly slow, and in this I agree heartily.

At this point, let me refer briefly to my own experience, starting as Secretary of Agriculture. Naturally, if our government could refine the moisture pattern of the United States, the Department of Agriculture could be very useful in helping farmers meet whatever goals might be set for their various crops. So when this cloud-seeding project was announced, I saw in it not only the possibility of immense benefit to farmers but also the chance for harm. A rain which helps one segment of agriculture may be disastrous to another. Hail destroys crops wholesale. The milking of moisture from clouds over New Mexico might produce drought in Kansas. I became interested in the subject and examined the possibility of measures designed to understand, encourage and regulate rainmaking. By then the subject had attracted many other followers, including the National Science Foundation.

Later, in 1949, after I had become a member of the United States Senate, I tested my ideas as legislative theories. There were many senators and representatives—and I was one of them —who introduced legislation aimed at controlling commercial rainmaking at least until the processes were better understood and their impacts more clearly defined. The result was a weather control study, but the going wasn't easy. People experimenting in rainmaking did not wish to be disturbed.

Perhaps I should illustrate. One day a rancher driving into Albuquerque saw enormous black clouds forming above the Sandia Mountains which lie to the east of that city. They were seemingly filled with rain. How could it be induced to fall? The rancher had been talking a great deal about the possibilities of cloud seeding; so he engaged an airplane, filled it with as much dry ice as it could carry, got into the plane with the pilot, flew into the black clouds and gave them a heavy seeding of dry ice. To his amazement the clouds seemed to dissipate, and no moisture at all fell either on Albuquerque or his nearby ranch.

Later the ranch operators in the northeastern part of New Mexico arranged with a cloud-seeding firm to set up silver iodide generators on the headwaters of the Rio Grande in southern Colorado. The snow pack in that area is a source of our Rio Grande water for summer irrigation. The seeders would wait until a large snowfall seemed to be imminent, and then seed it in the hope that the snow would actually fall 100 or 200 miles to the southeast. Many of us felt they were reducing our normal snowfall because of improper seeding.

To take a third example, we were very much encouraged by the fact that there seemed to be a pattern of periodicity in the rainfall in certain parts of the country when General Electric was conducting Project Cirrus. General Electric would seed above the Sandia Mountains at Albuquerque every seven days. There seemed to follow a pattern of rainfall every seven days in an area from, let us say, Louisville to Buffalo, which lay in the normal path that the winds would follow across the country. At Omaha there was also some variation in the level of the tropopause, the dividing line between the inner atmosphere and the stratosphere below which nearly all clouds are formed. We felt that these phenomena might have had an association with what was taking place in the seeding of silver iodide crystals at Albuquerque.

However, just when we began to notice some periodicity in the weather pattern, the cloud-seeding business got going at high pace and the pattern of periodicity began to change. Some of us believed the reason to be that ranchers had acquired these silver iodide crystal generators and had placed them in their ranch areas. They would tell the cowpuncher or the sheepherder, who might not have had any scientific education, that as soon as the clouds got black, the cooker should get hot and the silver iodide should be fed rapidly into the clouds.

I do not know how many cloud formations were broken up by this random seeding in the late 1940's or how much actual rainfall was produced in other parts of the country, but apparently the value of the experiment conducted by General Electric was diminished by what had happened in the cloud-seeding pattern. That led to the suggestion of a controlled national experiment, in which all cloud seeding would be prohibited except that which was done by the federal government or which could be closely monitored and watched with the aim of careful evaluation in all parts of the country. Only by such a controlled experiment, many of us thought, could any final proof be obtained as to the value or lack of value of cloud seeding.

However, it was not to work out that way. Those people who were interested in seeding hired legislative representatives to present their side of the case. Communities which were already engaged in seeding wanted no interference with their practices and policies and, therefore, urged us to abandon our attempts toward legislation to provide a controlled experiment and substitute instead a study.

As an alternative, Congress authorized and funded in 1953 a Weather Advisory Committee to study and report to the President on the state of the rainmaking in the United States. The report was published early last year and its conclusions and recommendations widely publicized. It was agreed that more scientific study is needed; that we know very little about the weather process.

This report, together with the Congressional hearings and reports on other legislation dealing with weather modification, document the weather control efforts in this country. As yet there is no over-all control over rainmakers although fourteen states have passed weather modification laws of varying effectiveness.

Soon the report of studies conducted during the International Geophysical Year will be available. They are expected to confirm much of what is known and add to the general fund of weather knowledge. However, some of the findings may be quite new. I noted in a recent press announcement, for example, that the northern and southern hemispheres have been shown not to be separate weather identities as commonly believed in the past. The I.G.Y. studies have indicated that while the storms of each hemisphere may not cross the equator, the energy releases of weather in one hemisphere may and sometimes do invade the other with potentially significant consequences.

Many of us were prepared for that assertion. That is because the atomic bomb came under detailed global study during the same interval in which the weather process has been under new scrutiny. While these studies may not be outwardly associated, they are parallel and, in the end, may serve the same goal.

Greater knowledge, but not necessarily greater understanding of atmospheric movements, began with the detonation of the first atomic bomb. With the loosing of deadly fission particles into the atmosphere, scientists began to study air movements and the mechanics of precipitation with new vigor and with some apprehension.

Their effort was expanded several times following the explosion in the Marshall Islands in 1954 of the first thermonuclear device. That explosion not only vaporized a small coral island; it ruptured the troposphere, or inner layer of the atmosphere, spewing into the stratosphere countless tons of radioactive debris. We have proof that this debris is being precipitated in both hemispheres.

The consequences of that 1954 blast and its successors upon the atmosphere, upon the mechanics of weather—and likewise upon all living earth creatures—is as yet only dimly perceived despite considerable and conscientious efforts.

There are those who compare such a thermonuclear eruption with the notable volcanic eruptions of history. Others say radioactive particles produce ionization of air at extreme altitudes and that such artificial ionization may supplement—or interfere with—the role natural ionization plays in the natural weather-making process.

Others say there is no proof. So the question stays alive until greater knowledge and understanding are available.

In any event, the arrival of world-wide fallout from hydrogen weapons tests has given to the meteorologist hitherto unavailable tools for the study of the atmosphere.

Recent rocketry strides also are yielding abundant new information about the upper atmosphere.

The sum of all these developments is to bring home to the man on the ground—or, if you please, the man in the street—a new awareness of the weather. Just as it may have seemed that something was at last to be done about the weather, along came an invisible radioactive rain to invade our bones, affect our genes and thus tamper with our evolution. This broadens any discussion of the social, economic, and political implications of weather modification.

Unfortunately, in this broader field, I cannot resort to the measurement of microns as can Dr. Byers in his discussion of the scientific aspects of weather modification. I must measure multitudes, and I am painfully aware of a shortage of yardsticks.

Nonetheless, I believe the implications of world-wide atomic fallout have cast at least a dim finger of light down the path we must follow in assessing the implications of weather changes wrought by man for his own purpose, convenience, and reward. And since we are again approaching the unknown, I would like to invite your attention to a statement by an eminent scientist and mathematician, the late John R. von Neumann.

In *Fortune* Magazine for June 1955, in an article entitled "Can we Survive Technology?" Dr. von Neumann said:

> Present awful possibilities of nuclear warfare . . . [among them world-wide fallout] may give way to others even more awful. After global climate control becomes possible, perhaps all our present involvements will seem simple . . . once such possibilities become possible, they will be exploited.

Edward Teller, who has been called the father of the H-bomb, told the Senate Military Preparedness Subcommittee in November 1957:

> Ultimately, we can see again and again that small changes in the weather can lead to very big effects . . . .
> Please imagine a world in which the Russians can control weather in a big scale, where they can change the rainfall over Russia, and that—and here I am talking about a very

definite situation—that might very well influence the rainfall in our country in an adverse manner . . . .

What kind of a world will it be where they have this new kind of control, and we do not?

Thus we have two of the leading scientists of this age looking upon climate control as the "ultimate weapon." Many military experts may regard the ICBM as the ultimate weapon, but weather modification may well be. Weather warfare could be so applied that the operator of the modification could not only damage his opponent but escape undamaged himself—something that might not be said of multi-megaton thermonuclear blasts.

These are very real sociological and political implications of weather modification. The economic implications are inseparable from the first two. Who needs to be told how long it would take to bring the United States to her knees if it lay in the power of another country to deny to us at will our drinking water or our wheat crop, to alternately freeze us or burn us up, to flood our cities and scorch our farms? Pestilence and disease would walk arm in arm with such a diabolical weapon.

The effect of toying prematurely with such overwhelmingly vital and elemental forces would wreak havoc with the interwoven complexities of life on this planet, upsetting the careful balances built through all of evolution between the devourers and the devoured, between plants and animals, health and disease, flood and drought, freezing and thawing. Against such forces, existing social, political, and economic forms are no more substantial than the morning mist. Against such forces not even space offers succor until man is able to migrate from Mother Earth.

Are we in America prepared to deal with such a weapon? For that matter, do the United Nations possess a capability in this field? I doubt it, and that doubt impels me to suggest that a high order priority might well be assigned to the study of weather control. The side effects of weather modification may be the last of our discoveries but the first to which we should have the answers.

So I close by quoting again from Dr. von Neumann's article, "Can We Survive Technology?":

> The one solid fact is that the difficulties are due to an evolution that, while useful and constructive, is also dangerous. Can we produce the required adjustments with the necessary speed? The most hopeful answer is that the human species has been subjected to similar tests before and seems to have a congenital ability to come through after varying amounts of trouble. To ask in advance for a complete recipe would be unreasonable. We can specify only the human qualities required: patience, flexibility and intelligence.

To his fine words, I add my own hope that our patience, flexibility, and intelligence may prove equal to the task.

EDWARD A. ACKERMAN

Weather Modification and
Public Policy

DURING THE almost thirty years since I took my first course in meteorology this science, along with others, has experienced some remarkable changes in outlook. Not the least among the changes has been the recognition that meteorologists, to paraphrase Mark Twain, not only can *talk* about the weather, but they also can *do* something about it.

One feature has not changed: the professional problems of

EDWARD A. ACKERMAN was director of the water resources program of Resources for the Future from its inception in 1954 until the 10th of October last year when he became deputy executive officer, Carnegie Institution of Washington. He was instructor of geography, Harvard University, 1940-43, assistant professor 1943-48; professor of geography, University of Chicago from 1948 to 1955. During and after the war he fulfilled a number of federal government assignments while on leave from his university posts. He was chief of the Geographic Reports Section and Topographic Intelligence Subdivision, Europe-Africa Division, Office of Strategic Services, 1941-43; technical adviser, Natural Resources Section, General Headquarters, Supreme Commander Allied Powers, Tokyo, Japan, 1946-48; regional analyst, Hoover Commission, 1948; chairman, Committee on River Program Analysis, President's Water Resources Policy Commission, 1950; chief of the Natural Resources—Public Works Branch, U. S. Bureau of the Budget, 1951-52; Assistant General Manager of T.V.A., 1952-54. In 1957 he was consultant to the Advisory Committee

the meteorologist still are formidable, and his situation often is uncomfortable. Who among the professions is more coldly and more immediately faced with the social results of prognostication? And yet who is more constantly called upon for prediction? As never before he should have our active sympathy, for now he must analyze weather-maker as well as weather! It is no wonder that many meteorologists show a tendency toward professional conservatism.

Professor Byers has presented an appropriately conservative view of weather modification. Summarizing his statements in a few words, I should say that the vista before us is one of awesome potential, but little is proved thus far. He does not endorse (or refute) the conclusions of the Advisory Committee on Weather Control; and he limits his reporting of positive results from weather modification operations to dissipation of "thin, stratified layers of undercooled clouds" by dry ice or silver iodide, and the initiation of rain in small tropical cumulus clouds by water spray.

If I have any difference with Professor Byers, it is in my attaching more significance to the report of the Advisory Committee than he seems to. Until its conclusions are disproved or altered, that report offers a reasonable basis for water-use planning in the West. It gives a far better basis for immediate planning than no conclusions at all. But this does not detract from my complete agreement with the most important statement in his paper. He notes that application of weather modification has outrun the knowledge of what happens. "In cloud physics," he says, ". . . we need desperately to go through some of the beginning steps."

Insofar as there is evidence to use, I shall discuss the social implications of this situation. There is little to grasp that is firm. But through the gap in the curtain we glimpse potentialities of such sweeping promise that action is called for, now.

---

on Weather Control. Mr. Ackerman was born in Post Falls, Idaho, in 1911. He received his A.B. (1934), A.M. (1936), and Ph.D. (1939) degrees from Harvard University.

What is the promise? First there is the promise that the precipitation over mountain regions may be increased by cloud seeding. It may be increased enough to permit a moderate expansion of economic production in the mountains themselves or in adjacent regions dependent on their water. Next there is the promise that some of the economic losses that weather brings to man may be lessened. Cloud modification potentially has meaning in the struggle against floods, crippling harbor and airport fogs, excessive local snowfall, local air pollution, hail damage, ice storms, tornadoes, and lightning-caused forest fires. We suspect, although we cannot yet even estimate with confidence, that the weather-caused losses through direct physical damage, or from avoidable costs of operation amount to more than a billion dollars yearly for this country alone. We have no estimate, or basis for estimating, their costs in health and social dislocation.

These are the more modest and the more immediate potentialities. Beyond them are the promises, or, better, the possibilities of weather modification which may affect entire regions, continents, or even hemispheres. Professor Byers has mentioned Dr. Wexler's speculation about the modification of hemispherical radiation losses. This hypothetical modification is reasonable enough so that we must consider it no longer science fiction, but illustrative of events that we should be wise to prepare for.

We must decide not only what we should like to do about climatic modification ourselves. We also must be prepared for what others may wish to do, and what they eventually may be able to do about it. We should never forget that on our northern hemisphere there is the greatest land mass of the earth, three times the size of North America. More than half of this supercontinent is so dry or so cold that it has a very low productivity. When we consider that the productive half supports four-fifths of the human race, we realize that climate modification will play for very high stakes there. The stimulus to technical experiment in climate modification will be great, and the pressure for application of a technical discovery even greater. Furthermore, we must assume that the capacity to develop and exploit such technology may be present in Europe and Asia.

Suppose that a feasible means for modifying the qualities of the polar air masses is developed. Suppose that this technique can extend the growing season of Siberia, Mongolia, and Northern Europe by a month as compared with the present. Suppose further that it will have the by-product of extending the seasonal rains of North Africa into the central Sahara, and causing the failure of the winter rains on which the European Mediterranean depends. If such modification could be perfected, there is little doubt that its application would be attempted.

Almost certainly a technique which would change the climatic pattern of Europe, Asia, and North Africa would change that of North America also. Do we know enough about the planetary atmospheric circulation to determine our national attitude toward the application of a technique of the potentialities I have described? Its application might be entirely for peaceful ends. It is not inconceivable that the nation or nations controlling the modification technique would offer compensation, or even resettlement, to those affected unfavorably. We should be in a most difficult diplomatic position, particularly if our reaction had to be based primarily on emotional nationalism.

I have indulged in this brief speculation because I believe that our foremost current problem in weather modification is one of imaginative public-policy formation. The exploration of outer space has captured our minds, and our funds. However, it does not yet offer the same wide vista of altering the conditions of life for the human race as manipulation of the atmosphere. This nation should begin to treat this subject methodically and adequately before another scientific "crash" program becomes necessary.

From the point of view of public policy, what would methodical and adequate treatment of the subject mean? To me it means: assurance that we have the professional direction and the professional evaluation to plan and carry through a vigorous, imaginative program; the attraction of additional competent scientists to meteorology and other atmospheric research; consistent and adequate public financial support; and agreement about what specifically should be supported by public funds.

I believe that we already have the professional leadership and professional evaluation to commence a methodical program. If we do not yet have a specific program, we have the capacities for drafting a good one, rapidly. This has been shown in the Berkner Report to the National Academy of Sciences on meteorological research and education, in some current federal programs, and in other ways. I also regard the second problem with some assurance for the future: that is, the attraction of additional competent scientists to these fields. If we have adequate financial support, and a well drafted, stable program of research, the scientists would be attracted over the long run. In this field, however, they cannot be counted on to appear spontaneously in response to calls for a crash program.

Consistent and adequate public financial support is more of a problem. Dr. Thomas Malone in a National Science Foundation lecture (January 1959) noted that we currently spend about ten cents per year per employed worker for basic research in meteorology. This is not adequate. Surely we could afford ten times that amount for fundamental research on the atmosphere. We could do this even if we had to organize some other public activities more efficiently to release the needed public funds.

Dr. James Killian, in an address before the American Association for the Advancement of Science, December 30, 1958, described meteorology as a field "where additional capital funds and emphasis are necessary." The President's Science Advisory Committee's report of December 27, 1958, "Strengthening American Science," mentions meteorology as a field suffering from "serious underinvestment." This problem, then, may be on the verge of receiving attention, but as yet we do not know when "additional funds" will mean "adequate funds." We do know, however, that greater support for meteorology will be very seriously considered in future governmental science budgets. Support must be increased to the point where the study of atmospheric processes is recognized as one of substantial future importance in our national scientific program.

There remains the last problem, that of agreement about what specifically should be supported by public funds. In my view this is a key problem.

In a very tentative manner, and for discussion, I should like to suggest the desirable content of a public program on meteorology and weather modification. The ingredients of my proposed program rest on: (1) distinguishing between fundamental and applied meteorology, or between weather *formation* and weather *modification,* if you prefer; (2) distinguishing between the dynamics of planetary circulation and cloud physics; and (3) recognition of the close relation of economic studies to the progress of weather research and weather modification.

There would be two foundation stones to my hypothetical policy and program. The first is the establishment of two distinct programs: one to include cloud physics, weather formation, and weather modification; the other to study the dynamics of the atmosphere on a planetary scale, and *climate* modification from now speculative manipulation of the planetary circulation. For brevity, I shall call the first the *cloud-physics* program, and the second, the *planetary program.*

The second foundation stone of policy would be clean separation, in program, of fundamental and applied science. Meteorology in this country has a history of mixing the fundamental and the applied aspects. Although this may be beneficial on occasion, in meteorological work it has led to a confusion of view, and a confusion of public objectives. As Professor Byers notes, the organization and pursuit of fundamental research in meteorology has suffered thereby—one might add, to the frustration of our present desire to shape weather to our needs.

The two programs might be broadly conceived somewhat in these forms:

The cloud-physics program would be concerned with all of the relatively small-scale and "meso-scale" phenomena of weather formation. Thus the dynamics of clouds of all forms, micrometeorology and microclimatology, and the dynamics of systems like hurricanes, would be treated. This would be a program of fundamental research, with both theoretical and experimental scope. Provision would be made for stable long-range support of scientists interested in theoretical studies, with complete freedom of individual investigation. This program also would plan, and, where necessary, carry out, specific programs of observation essential to the orderly research. It would include experimental work on any appropriate scale. For example, there must be fascinating experimental possibilities of a scale between the laboratory cold box experiment and tampering with actual clouds in the atmosphere itself. Such a program would be mindful of a comment in the British publication, *New Scientist* (December 25, 1958): "In the progress of science, random measurement is no substitute for well-conceived and economical experiment." We should consider seriously administering such a federal government program separately from existing meteorological agencies having heavy commitments to applied meteorology. They include agencies responsible for both forecasting and defense applications. With such separation we should have the best chance of producing a program with the essential balance, depth, and originality.

What would we do with weather modification, or cloud modification, as it now is practiced? As far as the federal government is concerned, I should suggest leaving it alone at the outset of this program. Defense and forecasting applications could stay as they are. The remainder, cloud seeding as practiced for the last few years, I should leave to commercial or private operators, to be regulated in their operation by the individual states. Until we have the understanding of the atmosphere that the cloud-physics program would eventually provide, further development of weather-modification techniques is more like seeking a prospector's lucky strike than an orderly program of exploration and discovery. Since we do not understand weather formation we throw assorted materials into assorted clouds, to see what happens, if anything. Even if a policy of federal government abstention from this field should result in a Supreme Court case or two, the federal government can contribute far more to weather modification through the cloud-physics program than in any other way.

The planetary program would require us to stretch our minds much more. I would keep before us the Science Advisory Committee's admonition: "Somehow diverse proposals of extraordinary novelty and import must be generated." The long-range goal of the planetary program would be nothing less than a working understanding of the vast, complex hydrodynamic system which flows around the globe. At the beginning the program would be heavily weighted on the side of theoretical and observational works. Over the years, however, an experimental phase of now unknown extent and content is conceivable. In it we should be concerned with the modification not only of precipitation but also of temperature. This possible program would offer a real opportunity for, and might even require international scientific co-operation. Responsibility for it might well be placed on a professional planning and review board composed of the most eminent meteorologists and physicists willing to serve, whatever their country. We could envisage its soliciting research aid, or placing research contracts, in any part of the world where the subject might be treated best. We have an opportunity here to show scientific leadership, and in so doing we could take a step toward forestalling or at least mitigating later problems in diplomacy. What may seem visionary today may be tomorrow's scientific capital, tomorrow's weapon, or tomorrow's disaster.

Economic and social studies (which need not be expensive) would be essential adjuncts of both of these programs. I think that this necessity will be obvious if one question is asked: Do we know what kind of weather or climate we should like to have if we could change it to order? If the answer is not evident, try asking it of people in different walks of life. We are concerned here with a question of the deepest social complexity, touching even upon the mental and physical vigor of our people. Here the unity of the social and the natural sciences becomes apparent. It is not too early to examine the foundations of such a question. While the answers for local weather modification may be easily attained, those for climatic modification are far more obscured.

All of these suggestions leave unanswered many related questions. The relation to work in progress now, the place of the proposed National Institute of Atmospheric Physics, the relation to the World Meteorological Organization, the relation to the National Aeronautics and Space Administration, the relation to universities, and the security implications of such programs—would have to be reviewed in more careful planning, and in exploratory discussion.

I shall add a closing remark about motivation. I have emphasized the practical reasons for considering a changed public policy on atmospheric research and experiment. I also might have followed the President's Science Advisory Committee as it reported on outer space—we must accelerate research to maintain, or recover, prestige abroad. But I like to think that neither of these is the best motivation. Our firmest ground for action, and the surest road to final success, is a compelling, unselfish need to know—to know all that the application of reason can tell us. We might well follow the guiding words of an eminent present-day Englishman, Sir Henry Self: "[Man] is Nature's agent by the gift of his divine reason . . . it is his duty to develop and expand [his reason] until he can really come to grips with the scheme of things." If we are able to plan these scientific programs in terms of the strategy of knowing, then inevitably all else must follow.

iii

EXPLORING FOR MINERALS

John A. S. Adams:
NEW WAYS OF FINDING MINERALS

James Boyd:
THE PULSE OF EXPLORATION

Paul W. McGann:
ECONOMICS OF MINERAL EXPLORATION

JOHN A. S. ADAMS

New Ways of Finding Minerals

HOW IMPORTANT is it to find better methods of exploring for minerals? For the long pull, no one can say for sure. Perhaps future generations will be most concerned with radiation levels or with population pressure on food supply and living area, and will look upon our efforts to anticipate their needs as totally unnecessary, although well meaning and even quaint. On the other hand, it may be that mineral shortages will limit mankind's ability to provide better for more people, so that some future generation may curse us for leaving them an earth depleted of mineral resources. The fact that presently-known

---

JOHN A. S. ADAMS, geologist and geochemist, has been associate professor of geology at The Rice Institute, Houston, Texas, since 1954. He was project associate geochemist, Department of Chemistry, University of Wisconsin, from 1951 to 1954. He is a Fellow of the Geological Society of America, and a member of the council of the Geochemical Society. Mr. Adams received a Fulbright grant to the Geological Museum in Norway, 1949-51, and was a member of the Norwegian expedition to Svalbard (Spitsbergen) in 1950. He has done other field work in Mexico, Central America, much of the western United States, and, to a lesser extent, in Europe. He was formerly consultant to Shell Development Company, and at present is consultant to the Humble Oil and Refining Company. He was born in Independence, Missouri, in 1926. He received his B.S. and Ph.D. (geology) degrees from the University of Chicago.

domestic reserves of many important minerals can be exhausted in twenty-five years or less would make it seem that mineral shortages are inevitable. From today's standpoint, therefore, the supply of minerals, and better techniques of enlarging them, are practical questions of some urgency. Adjustment to any plausible physical limit can only be made by striking an ultimate balance between supply on one hand, and population and per capita consumption on the other. To calculate this balance, however, it is essential to know the limit of the mineral resources for the future.

Future mineral resources will be determined by several mutually dependent technological factors. These basic factors include: (1) synthesis of minerals; (2) development of substitute materials; (3) new or greatly expanded demand for particular materials; (4) development of economic techniques for using lower and lower grade ores even down to the parts per million found in the common rocks; (5) new and more extensive exploration for high-grade mineral deposits. The last factor—mineral exploration—is the main subject of this discussion, but it is necessary to consider briefly how these other technological factors can change, or perhaps even eliminate, the need for mineral exploration.

Many minerals are used in the form in which they come from the ground. Building stone, sand, clay, and limestone are in this category. In all but a few cases these minerals are so abundant that their large-scale synthesis will never be necessary. There will be a shortage of standing room on the earth before there is a shortage of granite. Some minerals, however, such as industrial diamonds, have been so rare and so essential that their synthesis has been a long-sought technological goal. An important and almost final step toward that goal was the commercial synthesis of industrial diamonds.

Synthetic diamonds have been available for nearly a year at a price of about four dollars a carat. This price is almost 40 per cent higher than that of natural stones; the synthetics command a premium because they are more uniform. Well over 100,000 carats have been synthesized and current production capacity

is enough to supply about a fourth of current domestic demand. I presume that the process is economic and that General
Electric will recover the two and a half millions spent on
research and the additional sums spent on expanded plant. If
they do not, we can predict with confidence that there will be
some trouble between the stockholders and management of
that company.

The synthesis of industrial diamonds affects mineral exploration in several ways. At no time in the future need there be a
shortage of industrial diamond dust; the only remaining industrial advantage of natural diamonds lies in those few applications requiring the larger stones. Even this advantage may
be eliminated if larger synthetic diamonds can be made or if
the present synthetic diamonds can be made to do these jobs.
Today, in 1959, any economic diamond deposit discovery will
have to be larger and of higher grade than it needed to be
even three years ago. These are some of the implications for
diamonds as an *object* of exploration. There also are implications for diamonds as an *instrument* in exploring for other
minerals. To the extent that synthetic diamond prices decline
—which would be the usual pattern—the costs of diamond bits
used so extensively in mineral exploration may be expected
to decline.

In considering the future, one must also remember that the
synthesis of diamonds was anticipated. For example, in 1952
the President's Materials Policy Commission stated that substantial progress in the synthesis of diamonds could be expected
by the 1970's; certainly this expectation has been amply fulfilled well ahead of schedule. The Commission also anticipated
the substantial progress that has been made in synthesizing
other critical minerals like electronic grade mica.

The recent developments in synthesis are revolutionizing
mineral exploration. The problem is becoming less and less a
matter of finding a certain element like carbon in a certain
rare form such as diamond. The main problem in mineral
exploration is becoming one of finding economic sources of
the chemical elements. Even if a few minerals, such as optical

grade calcite, are never synthesized economically, there are still adequate or superior substitutes.

Synthetics and substitutes are the twin children of chemical technology. The same researchers who achieved the commercial synthesis of diamonds also made the first cubic boron nitride ever seen. It had long been suspected on theoretical grounds that cubic boron nitride should be as hard or harder than diamond. It turned out to be harder, and in this respect cubic boron nitride is a super diamond. Who is to say that the manufacture of this formerly unknown material may not reduce or even eliminate the demand for industrial diamonds? To cite another example, already the substitution of transistors for electronic tubes means that electronic grade mica is not to be found in every modern radio.

Although synthesis and substitution eliminate some of the problems of mineral exploration, they also create new and unusual demands. Thus, the development of transistors created an unanticipated and greatly expanded demand for germanium. Our recent history is full of such examples. Who would have predicted twenty years ago that within five years the military security of the country would depend upon a vast expansion in uranium mining, or that the demand for fluorine and lithium would triple and continue to grow after the war? Considering automation and space research, who can predict the mineral demands of our technology in the next few years?

One can only say that the long-term demand for all elements will grow, but some demands will increase much faster than others, and we will not necessarily be looking for the same ores and other minerals that have served mankind in the past.

But, allowing for much greater flexibility, we and future generations still must rely upon mineral exploration and the development of economical extraction techniques.

Over the past fifty years there has been a steady decline in the average grade of domestic ores discovered and mined. This decline in grade has been generally offset by more efficient methods, including mechanization, for extracting metals from these ores. By such means the long-term real price of many metals like copper has actually been reduced. Important and sometimes unexpected advances in extraction technique continue to be made. For example, the helium wells around the Panhandle of Texas are unique geologically and for conservation purposes they have long been under the control of the United States Bureau of Mines. Extensive drilling for oil in many parts of the world has not uncovered anything like this source of helium. The Bell Laboratories have recently announced the development of an economic method for extracting the small amounts of helium long known to be in natural gas. This promises to relieve greatly the drain on one of our most limited resources. But is this merely another example of careful and skillful scraping of the barrel with extraction techniques becoming less and less effective? There are some indications that exploration and extraction techniques are not improving fast enough to keep real costs down. Even if this is true, does it indicate a lack of scientific and technical ability and power to develop better extraction and exploration techniques?

Scientific and technical power of this type is analogous to naval power as described by Sir Winston Churchill in *The Gathering Storm,* the first volume of his *The Second World War.* During the first year of World War II, Sir Winston, after being questioned about some German successes with their surface fleet, said:

> When we speak of command of the seas it does not mean command of every part of the sea at the same moment or every moment. It only means that we can make our will prevail ultimately in any part of the seas which may be selected for operations, and thus indirectly make our will prevail in every part of the sea.

Similarly, a great potential in the development of mineral exploration techniques does not mean necessarily that every ton of ore has been surveyed out for all time. It means that, given a few years and adequate support, foreseen and even unanticipated demands for minerals can be met, even if superior substitutes and more efficient extraction methods are not developed. Uranium provides an excellent illustration.

In 1945 the proven uranium reserves in this country were so low that all reserve estimates were classified to conceal what could have been a great political and military weakness. As a matter of national policy, a strenuous effort was made to develop domestic reserves. By 1954, these ore reserves were measured in millions of tons. On November 24, 1958, citing a need to prevent overproduction, the Atomic Energy Commission withdrew its announced intention to buy all ore developed and delivered after 1962. The Commission stated that it did not anticipate any reduction in its needs, and domestic ore reserves of about 80 million tons were publicly proclaimed.

Whether or not the ore was discovered in an economic way and whether or not eight dollars a pound is a fair price or an open market price are complicated questions best left to the economists. There may be no more agreement about these costs than there is about the costs of electricity from the TVA. For the present discussion it is sufficient to observe that large reserves of ore were discovered and the cost was not so great that it affected noticeably our standard of living or our military budget. Furthermore, raw uranium at eight dollars a pound represents an inconsequential fraction of the cost of a nuclear weapon or a nuclear power reactor.

This great potential in mineral exploration was realized with uranium because the federal government provided the economic incentive and backing to mobilize a substantial fraction of the professional personnel and facilities available for exploration. The organizations built up may have almost approached the numbers and equipment that the petroleum industry devotes to exploration operations and research. Research on uranium exploration produced a number of new and improved techniques. These included better radiation detectors, better search patterns and techniques for both aerial surveys and exploratory diamond bit drilling, methods for finding clues to uranium deposits in plants and natural waters, and many others. However, the majority of the geologists and other scientists supported in this effort by federal funds were employed in applying the techniques, particularly in helping the private prospectors use the available techniques and knowledge.

'I'he private and largely amateur prospectors, numbering in the tens of thousands, supplied without cost to the federal government most of the man-hours spent in locating the uranium deposits. The search for uranium was turned into a kind of national lottery, with Geiger counters serving as tickets, and discovery and production bonuses serving as prizes. This was a very popular lottery and the rewards of uranium discoveries were very much in the public mind, pervading every part of our culture, including the comic strips. As in all lotteries, only a very few won anything. However, their success was not solely a matter of luck. At the very least, the few winners had mastered the use of an instrument like the Geiger counter; at the very most many had educated themselves by reading everything available on the subject and had taken every advantage of the available professional advice. A very small percentage of the prospectors were killed; flying light planes down canyons to make radiation surveys is a tricky—though commonly successful—exploration technique. As in all lotteries, some people went broke. However, the great majority of the private prospectors neither won anything nor suffered any great physical or financial hurt. I believe that most of them enjoyed tramping the hills, that searching for uranium provided many with amusement, adventure, or even a serious and patriotic purpose during their week-ends, vacations, and retirement. The private and largely part-time amateur prospector demonstrated in the uranium exploration program that if he is given sufficient incentive he will buy his own exploration equipment, learn to use his equipment in an adequate and even expert way, and spend countless hours exploring with that equipment. We must consider the possibility that the use of the private and particularly the part-time amateur prospector on this scale represents a new and major technique for mineral exploration in the national interest. With uranium only two things were required: relatively modest expenditures by the government for bonuses, and the development of the search into a sort of popular lottery or fad.

In citing the recent search for uranium as an example of what is technologically possible in mineral exploration, it is important to consider whether or not uranium is in some way unique, relative to other essential elements. In uranium exploration, ore can be detected with a radiation instrument. Assuming the instrument functions properly, potassium and thorium are the only other elements that give a similar effect; at certain places and times cosmic radiation and radioactive fall-out may also confuse the uranium prospector. However, as far as the individual prospector on foot is concerned, many metals can be detected by means that equal or surpass the Geiger counter in economy, speed, simplicity, and reliability. In recent years the United States Geological Survey has developed a number of simple chemical tests for metals such as zinc. One commercially available device that should delight the gadgeteer is a portable detector for beryllium that uses a neutron source to make some of the beryllium, if it is present, radioactive; the weak radioactivity so induced is then detected. Such devices and tests make possible the detection of deposits that would have escaped the sharp eye of the old-time prospector, who relied mainly on color and sight identification of minerals. There is every reason to believe that for almost any metal, it is now technically possible to equip the prospector on foot as well, or better, than he was equipped for the uranium search. Some of the methods are well developed, others would require—relative to the uranium program—a very modest increase in research effort.

Exploration techniques for uranium are almost unique in the effectiveness of rapid and cheap radiation surveys from the air or ground vehicles. An equally effective exploration technique is the air-borne magnetometer which locates iron or nickel deposits by their magnetic effects. However, for the great majority of metals, there are no presently effective methods of aerial exploration. An intriguing and largely unexplored possibility lies in the use of color aerial photographs to detect slight peculiarities in the color of soil or plants that may be due to the presence of metals from underlying ore deposits.

But uranium is not unique in its average concentration in the earth's crust. As a matter of fact, its concentration is very close to the median; half of the known elements are more abundant than uranium and half of the elements are less abundant. Nor does uranium happen to concentrate in ore deposits more than other metals; actually uranium tends to be more dispersed and harder to find than many metals. In short, there are no overriding technical or scientific reasons to conclude that exploration for uranium is either markedly harder or markedly easier than exploration for most metals. It is only different.

A final point to be considered about uranium exploration is the ultimate potential of our present exploration techniques for uranium. At present, the exploration effort is being reduced because the Atomic Energy Commission no longer offers to buy all uranium produced after 1962. The exploration effort is not being reduced because anybody believes there are no more ore bodies to be found or because of a lack of competent people and techniques to locate new ore. Indeed, the exploration techniques have proven so increasingly effective that for the last year or so overproduction has been officially announced as the main short-term problem. This has led to a drastic reduction in the exploration support and staffs. In our society the exploration geologist frequently shares the fate of the professional military man. In time of need he is a hero, but after the need of society is met by skilled and devoted service, sometimes beyond all expectations, he is forgotten and often dismissed to suffer considerable personal hardship. Obviously, 80 million or so tons of proven domestic uranium ore does not represent the technical limit of what can be found. The technical limit is unknown, but the fact that more and more ore had been located at an ever faster and more efficient rate for almost fifteen years would indicate that there is still a great deal of ore to be found. We must also keep in mind that the intensive research in exploration methods was paralleled by intensive research in extraction methods from many types of ore, including such low-grade sources as phosphate rock. Thus uranium reserves could be increased greatly by further testing of the known and promising extraction methods for the large low-grade deposits already located.

In evaluating the domestic ore reserves of metals other than uranium, one must remember that with these metals exploration research and operations have been rather meagre compared to the uranium program. In cases like lithium and molybdenum, where there are adequate domestic reserves for many years, there is little reason for the mining companies involved to invest much money in exploration research and operations; they spend money on encouraging and developing new uses for their particular metal. In the cases of those many metals with low or nonexistent domestic reserves, few exploration geologists estimate that the technical limit has been reached. Many geologists agree with the position taken by Thomas B. Nolan, director of the United States Geological Survey, at the 1958 Resources for the Future Forum. In a paper called "The Inexhaustible Resource of Technology," Dr. Nolan said:

> I believe that the prospect of impending shortages or unsuitable supplies will continue to inspire the research and technical advances that will make it possible to resolve such problems well in advance of the doom we often are prone to foresee.

To some, unfortunately, such optimism may sound like the pure and naïve faith in science of those who make of science an omniscient and omnipotent god with white-coated priests. Such uncritical faith is hardly the basis for future policy. The optimism of Dr. Nolan and other scientists who inventory the earth rests on a factual basis. In estimating how long high-grade ores will be discovered and recoverable, the following facts must be considered:

The United States is a large area and there is considerably less than one professional exploration scientist for every hundred square miles.

Many promising exploration techniques have not been tried extensively.

For the most part our present exploration techniques only scratch the surface—the problem of finding ore that occurs only at depth has not been solved.

There are and will still be many unexpected discoveries; for example, France has increased her European oil production tenfold in the last ten years.

There are many new and exciting research approaches to consider; among these are the possible uses of cheap nuclear explosions in mining; the use of radioactive tracers to solve problems in extraction and detection; automation in instrumentation that increases tremendously the amount of data available; computers to handle the increased amounts of data, to provide accurate and fast inventory of reserves, and to calculate efficient search patterns and the way reserves increase as grade decreases.

Aside from such technical aspects, the public in general and economists in particular may be concerned about who is to pay for all the elaborate equipment and research, and may wonder if it will all prove worth while. The problem of who is to pay is one that our society in general must solve. The question of whether or not more research in exploration techniques would be profitable can, however, be given a qualified answer on technical grounds. Considering the success of the uranium exploration program, and considering how little has been done in exploration for other metals, it would be contrary to all previous experience for an intensive program to fail in developing larger and more economic supplies.

Let us assume for the moment that better exploration techniques, superior substitutes, and better extraction techniques can provide the minerals for several centuries, even including, perhaps, a long cold war. It is still possible that despite the best efforts, exploration technical development at some point will fail to locate sufficient ore. Such a judgment, however, could not be reached overnight. Only after the expenditure of substantial sums for some years would it be possible and fair to conclude that the technical limit of development may have been reached. Exploration for petroleum in this country may be reaching just such a point of diminishing returns for exploration effort.

For many years the petroleum industry has employed the majority of geologists, geophysicists, and other exploration and production scientists. Huge sums are spent on exploration equipment and laboratories that are unsurpassed by anything else in mineral exploration. The industry develops and applies new techniques quickly, as in the case of offshore exploration and drilling. Technical advances in other fields often find quick application in petroleum exploration; magnetic tape to record geophysical data is one example; computers and radioactive tracers in production research is another example. The industry also carefully considers such novel suggestions as the use of underground nuclear explosions in producing oil. Despite all of this, the cost of finding oil in this country has risen since about 1949, and although "wolf" has been cried many times about petroleum reserves, some, but by no means all, petroleum geologists conclude that the recent and slight declines in discovery rate will continue. Although there is no reason to doubt that exploration techniques will continue to improve, some students believe that a major or revolutionary breakthrough in exploration technique will be necessary before it will be possible to discover petroleum and gas reserves faster than they are consumed.

Even if there is no such major breakthrough, it does not follow that any failure in exploration will bring a sudden and terrible day of judgment when oil suddenly runs out or becomes twice as expensive. If there is no major breakthrough in exploration techniques for petroleum, exploration and production in this country will slowly decline over several decades. Alternate sources of hydrocarbons and fuel will be foreign petroleum and domestic solid sources. One of these domestic sources, gilsonite, a superior kind of asphalt now used chiefly in paints and varnishes, is already in commercial production.

The extraction of oil from the considerable reserves of oil shale is so close to being commercial that an oil company has tried to operate a pilot plant to determine precisely the economics. This plant is now closed, but there is little doubt that it will reopen at some time in the future. Coal and tar sands are other sources of hydrocarbons with considerable reserves. At worst these abundant hydrocarbons will be only slightly more expensive; at best they may be cheaper than domestic petroleum is today.

If these predictions for oil prove essentially correct, and if their broad implications apply to other minerals, the worst that we can conclude is that for many minerals the technical limit of exploration may be reached within a few centuries. Increasing ability to substitute materials will be very important. By-products of agriculture, forestry, and fisheries may have to provide an alternate and renewable source for hydrocarbons needed for plastics and synthetic fibers. However, if no alternatives to cheap power from oil, coal, and other mineral hydrocarbons are developed, some future generation will indeed condemn us for burning the fossil fuel candle at both ends.

At present, nuclear power and solar energy are two power sources that are technically feasible and becoming more economic every day. Either of these two sources is capable of providing energy in the tremendous amounts needed. The previous discussion already indicated the considerable amounts of uranium available for power generation. In addition to uranium, thorium, which is more abundant, and deuterium, which is still more abundant, are sources of nuclear power that can and will receive serious attention. Solar energy is equivalent to several barrels of oil per year for *each* square foot of the earth's surface. Future standards of living may be seriously lowered if nuclear power does not become cheap and abundant during the few centuries it may take to exhaust fossil fuels. It is important to note that the rate of technical discoveries in the use of nuclear and solar energy is increasing. It cannot be demonstrated that nuclear power or solar power will never be considerably cheaper than our present power. The people working in these fields would be the last to say that they have been so clever that no future improvements can be made. When mineral exploration becomes uneconomic, energy supplies and costs will clearly be the ultimate physical limits on the standard of living.

Assuming the economic synthesis of rare and useful minerals like the diamond, the situation when mineral exploration is no longer economic can be discussed in terms of the chemical elements. Of the first hundred elements discovered, sixteen are of little interest in this discussion because they are radioactive with only a very transitory existence, and they are the least abundant elements. They are either daughter products of thorium and uranium or they are man-made. Radium is the only one that ever found much use, and it is largely supplanted by cobalt-60. At present twenty-one other elements have little or no use, although efforts are being made to find uses for them. The remaining sixty-three elements are used in our present technology. Of these sixty-three essential chemical elements, exploration never began or has essentially ceased for eight of them. Silicon in the form of high-quality glass sands is available in almost every area; calcium in the form of limestone is so readily available that exploration and reserves will continue to be matters of minor interest. The first economic source of free oxygen, argon, and neon was the atmosphere; and exploration for these elements was never undertaken because it was clear that energy costs for extraction from mineral sources were prohibitive. At the turn of the century, mineral sources were the only sources of nitrogen. Today the atmosphere is the major source of nitrogen and mineral sources are minor and a large nitrate discovery might be a very marginal proposition. In the last fifty years the oceans have become the economic source for magnesium and bromine. If present techniques on brines become slightly more efficient, perhaps in conjunction with large-scale desalting of sea water, it is quite likely that the oceans will become the major source and reserve for sodium, chlorine, and potassium, as well as other elements.

The extraction of aluminum from clays and of iron from taconite would indicate that the day may not be far distant when exploration for ores of these metals may greatly diminish or even cease. Some of the elements mentioned are the most abundant on the earth, but others like neon are rare. Indeed, the least abundant of the stable elements is xenon, a very rare gas; for this element and the related krypton, there will never be any exploration, and the available and ultimate reserves in the atmosphere can be readily calculated. In calculating ultimate reserves of a chemical element one must allow for the discovery of new deposits and more efficient extraction techniques. For many elements it is very difficult to make estimates about costs, and new discoveries. However, in considering whether or not supplies of chemical elements could be limiting factors, it is useful to consider the concentration levels shown in the accompanying figure. The cosmos is mainly—perhaps

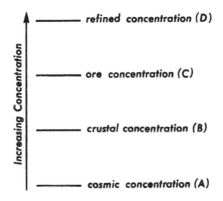

90 per cent—hydrogen and helium. Thus the cosmic concentration (A) of nearly all the other elements is well below the crustal concentration (B). The term "crustal" as used here includes all of the earth's surface environment, including the atmosphere. The main objective of mineral exploration is to

find places where natural geologic processes have concentrated the element up to ore grade (C). Extraction and refining techniques, using substantial amounts of energy, further concentrate the element to the refined level (D). In general the ore or economic level (C) has declined in recent years, as new ways have been found to profitably use lower concentrations. The refined level (D) has risen in many cases. Reactor grade graphite, reactor grade zirconium, transistor grade silicon, and phosphor materials suitable for television screens must have impurity levels measured in parts per million or less. Thus, even before fabrication, the trend has been and will continue to be toward the expenditure of more and more labor and energy on each pound of material.

Exploration effort diminishes or even ceases when the ore level (C) is pushed down to or below the crustal concentration level (B). This has already happened for the gases extracted from the atmosphere (oxygen, nitrogen, argon, neon, krypton, and xenon), for the elements extracted from the sea (magnesium, bromine, and in some areas calcium as oyster shell), and for silicon in the form of quartz sand. For most of the elements just listed, the reserves available with present energy costs and extraction technique are measured in tens or hundreds of pounds for *each* square foot of the earth's surface. In the case of xenon, terrestrially the rarest of the stable elements, there is about one pound available for each fifteen hundred square feet of the earth's surface. It is very probable that before the exhaustion of ore bodies, efficient extraction techniques will eliminate or greatly reduce the need for mineral exploration and make adequate reserves of many more elements available. A recent and intriguing development along these lines is the possibility of recovering manganese, cobalt, nickel, copper, and even thorium from nodules that occur on the deep sea floor.

To the extent that the elements can be recycled through nature or as scrap they can be used over and over like the oxygen in the atmosphere. The number of elements that may eventually be recycled depends partially upon the skill and ingenuity of extraction techniques. But even with the most ingenuous extraction techniques, it will be necessary to have cheap and plentiful energy sources. It is clear that if energy were to become expensive or scarce, there would be a need to go back to exhaustible mineral deposits for elements like nitrogen. On the other hand, with bountiful energy, lack of materials is unlikely to limit the number of people that can live as well or better than we do.

In conclusion, we can expect that for many years to come the best exploration efforts will be needed to provide many of the essential chemical elements. However, for several of these elements, we will probably live to see better extraction techniques, superior substitutes, or cheaper energy, relieve us and all future generations of dependence on the high-grade deposits found by mineral exploration. For ourselves and as a resource for the future, it would seem both prudent and profitable to intensify the research effort on exploration techniques so that we can make a more complete and useful inventory of the earth. If this is done, there is reason to believe that the retreat to lower and lower grade ores can continue to be orderly, cheap, and skillfully executed with mineral exploration providing more than ample time for the development of better extraction techniques, superior substitutes, and cheap energy sources.

JAMES BOYD

The Pulse of Exploration

FUTURE STUDENTS of twentieth century economics will be astonished to see how far political and economic theory went astray during this period of history through widespread misunderstanding of obvious facts governing supply and demand in the field of mineral raw materials. They will observe that shortly

JAMES BOYD, vice president in charge of exploration of Kennecott Copper Corporation since 1955, was previously (from 1951) exploration manager of that company. He was an instructor in geology, Colorado School of Mines, 1929-34; assistant professor in mineralogy, 1934-37; associate professor of economic geology, 1938-41; and dean of the faculty, 1946-47. He was director of the Bureau of Mines 1947-51, also Defense Minerals Administrator 1950-51; geologist with the U.S. Geological Survey, 1933-34; consultant in geology, mining, and geophysics, 1935-40; and president and general manager of Goldcrest Mining Company, 1939-40. He served as chief of the metals section, Office of the Under Secretary of War, 1941; and was director of the industry division, Office of Military Government for Germany, 1945-46. He has been chairman of the National Science Foundation Committee on Mineral Research since 1952. Mr. Boyd was born in Kanowna, West Australia, in 1904. He received his B.S. from the California Institute of Technology, and M.Sc. and D.Sc. degrees from the Colorado School of Mines.

The author wishes to express his gratitude to Julian W. Feiss for his helpful criticism, suggestions, and editorial assistance.

after the beginning of the Second World War the expanding consumption of minerals and their products soon outstripped the capacity of producers to satisfy the demand with existing facilities. Furthermore, they will note that the long-remembered experience of excessive capacity in the extractive industries following the First World War had made producers reluctant to expand capacity, and they will recognize that this was the main reason for the shortages following the Second World War.

Today, many of us recognize that mineral shortages that have occurred at various times since 1945 were largely of a temporary nature. However, there are others who have sincerely believed that these shortages were a prelude to approaching exhaustion of basic mineral resources. Herein lies a fundamental fallacy upon which far too many accepted economic theories and political actions have recently been based. Its imprint is clear, to name only one example, in many of the essays in *Perspectives on Conservation,* the book that resulted from the 1958 Resources for the Future Forum. I wonder whether Dr. Adams has fallen into this trap.

Granting that production problems of each separate mineral commodity differ, and that it is dangerous to generalize, if we dig into the problem deeply enough the fundamental supply factors are essentially the same for all producers. With a few exceptions, usable mineral materials are present in the earth's crust in quantities that are sufficient to defy any ordinary combination of social and economic events to exhaust them. This is not a wild assumption; it is a fact that has been demonstrated by scientific observation.

If we start from this premise, and not from the assumption that resources are necessarily exhaustible, we can encounter with greater equanimity the economic and politico-social problems arising from the maldistribution of economic mineral concentrations throughout the world. I believe that we should henceforth stop prattling about the exhaustion of basic resources and concentrate upon solving the problems of economic supply and distribution, for these are difficult enough without

the additional complexity of misconceptions. Once these misconceptions are overcome, we can begin to analyze minerals
exploration efforts realistically.

Before evaluating the specific economics of exploration, I
should like to point out two other rarely mentioned facts that
have an immediate bearing upon this problem. First, no mineral raw material has ever ceased to be of use because its supply
source has become exhausted. With few exceptions every new
application of a mineral or metal has stimulated and developed
new sources of supply providing, of course, it has proved economically feasible to produce and market the product. There
are few minerals, or their resultant metals used in industry or
science, for which substitutes cannot be found in one form or
another. If an economic use has been found for a material,
and it has been needed badly enough, supplies have in time
of peace been made available through exploration and development. It is recognized that the more urgent demands for
defense have at times required the external stimulus of special
government programs, such as were used for the development
of adequate supplies of uranium, columbium, tungsten, etc.,
in recent years, but such stimulus is an emergency measure.

Second, ore bodies of major importance have an almost
universal tendency to grow in apparent extent as they are
mined: the original concept of their magnitude is invariably
an underestimate. Furthermore, many mining enterprises continue to decrease their cost per unit of contained metal through
the application of scientific management and new technology,
despite decreasing grade, higher stripping ratios, depth of work-
ings, and rising wages. It is not surprising, therefore, that ore
reserve estimates continue to increase despite accelerating rates
of extraction. This, however, cannot go on forever and obviously some day existing mines must be replaced by new ones.

In view of these favorable factors in minerals economics, one
might ask why we bother with exploration at all. The answer,
of course, lies at the base of all present and future economic
development: despite their total abundance, mineral deposits
sufficiently concentrated to be economic at a given time and

place are rare and are distributed without regard to man's preference and convenience.

In addition to the materials supply problem there is the human problem. The all-pervading and inexorable trend of human events exists as the universal base of all future economic factors governing the growth of the mineral industries. There exists the inherent desire of man to improve his lot as reflected in much of the world's unrest, particularly within the under-developed nations. If these desires are to approach even 10 per cent of the living standard within the United States, then there is not enough productive capacity currently in existence or in contemplation throughout the world to meet the raw materials requirements of hoped-for industrial development. Yet it certainly is not too much to anticipate that this limited objective can be achieved in the not too distant future. Over the long range requirements for mineral raw materials have an unhappy habit of increasing faster than human concepts for their use. Today there appears to be ample capacity to meet the needs for most minerals during the next decade. But with the more troublesome commodities, such as copper, lead and some of the ferroalloys, etc., *economic* deposits upon which to plan expansion are not always known at today's prices and costs. I emphasize the word "economic" for this now becomes the primary consideration.

Dr. Adams discusses certain technological developments which might eliminate the need for mineral exploration. This discussion is unconvincing for the following reasons: If a mineral resource exists or has existed in sufficient concentrations to permit extraction at economic costs, there is no reason to believe that similar deposits which can be mined at similar cost levels cannot be found. He specifically refers to industrial diamonds, but it must be remembered that diamonds have for many years been subject to the price control of a cartel—an artificial price that would undoubtedly break if all diamonds were not subject to controlled marketing. To date, modern exploration techniques have not been applied seriously to diamond search and when additional diamond deposits are eventually found, there is no reason to believe that they will prove any less economic than those already operating. The fundamental question in regard to synthetic diamonds is not that of their competitive position against an artificial price for natural stones, but rather the question of their ability to compete on an open market—if such were to exist.

The exploration scientists sometimes have been accused of being too optimistic in believing that the application of technology can solve the problems of finding new mineral deposits. Here lies the basis of another misconception of what has happened in the past few years. Let us look at the record. Twenty-five to thirty years ago, virtually all new petroleum discoveries were made on the basis of the interpretation of the geological conditions observable from the surface. At about this time the easily accessible portions of the earth's surface had all been inspected from this point of view and the industry was forced to apply a reasoned scientific approach to the exploration for new oil resources. As in most tight situations, necessity was the mother of invention; the necessary incentives were provided, and the industry tackled the problem through the application of scientific research and new techniques. This resulted in the development of exploration seismographs, gravity meters, and other geophysical devices and tools—a development that is continuing today. As a consequence the oil industry has been able to keep pace with the demand and will continue to do so—at least until the costs of discovery rise to a point where other fuel, such as that of oil shales, can compete and take a portion of the supply.

Many other mineral deposits, which have also been found traditionally from surface evidence, have been sufficient to meet the demands of a rapidly increasing industrial economy even though these deposits were not fully developed or at the production stage some years back. Concerted scientific geology and prospecting methods were not applied to the search for deposits of these minerals to any extent until the last decade. Since the war, and starting with the development of airborne magnetic surveys (instruments that evolved from military usage) the search for new mineral deposits by more scientific means has been increasing at a steady pace.

The better application of inductive geological reasoning, assisted by geophysical instrumentation, has resulted in the discovery of more mineral deposits in the past ten years than had been found in the previous quarter century. In this connection, it is interesting to note that Professor Mason has recently mentioned* lead as being one of those metallic raw materials for which we were unable to find new important resources. This is no longer correct, if it ever was. Geological reasoning, aided by geophysical surveys and systematic drilling, has resulted in the discovery of at least one new lead mining district of grade and potential size to rival the original southeastern Missouri deposits, which for many years were the world's principal sources of lead. These newly discovered deposits lie over a thousand feet deep with no surface expression or indication of their existence. More important perhaps is the realization that within the geologic setting in which these mineral deposits are found, there exists an enormous area for the search for new and presently unknown districts. There are other equally significant examples available from the record of the past ten years, but this is sufficient to demonstrate that scientific principles, if properly and intelligently applied, are becoming the key to discovery of new mineral resources and are largely eliminating the fear of over-all failure.

To have an effective and useful minerals supply program, the discovery and development of new mineral resources must be accomplished economically. The raw materials base of a growing world industrial society must be sufficiently economic in itself to permit and stimulate normal industrial growth. Furthermore, such a minerals program should be sufficiently flexible and of a magnitude to provide for future economic expansion well in excess of current requirements.

* "The Political Economy of Resource Use," by Edward S. Mason, a paper presented at the 1958 Resources for the Future Forum. Published, along with the other 1958 Forum papers, in *Perspectives on Conservation* (Baltimore: The Johns Hopkins Press, 1958) .

It is here that we encounter conflicting political philosophies as they influence the regulation of extractive industries in many countries. Past experience has demonstrated that some form of incentive is required to encourage the search for new mineral deposits because prospecting and mine development are recognized as a very risky business. Here again it is worth reviewing the record as to what has happened in various countries in recent years.

We can ask ourselves, why is exploration progressing at such a pace in Canada while it has been reduced to a minimum in the United States? The difference lies in the attitude of the respective governments. In Canada, the discoverer of a mineral property can assume the normal highly speculative risks of mineral venture with full knowledge that he has ample opportunity to reap the rewards of his enterprise if he is successful. If he should find a mine, he knows in advance that for three and one-half years after it is put into production it will not be taxed; and, furthermore, that he may sell his discovery without being subjected to the capital gains tax which he would encounter in the United States. Dr. Adams has used the uranium industry as a fair example of what could take place by the use of favorable incentives. There are some countries in the world that provide these or similar inducements, and exploration proceeds apace. There are, however, others where the governments are not sympathetic to permitting rewards for taking such risks, and, consequently, very little exploration is undertaken. We have only to look at the difference between the development of the petroleum industry in Venezuela by contrast with the situation in Brazil, Mexico, and Argentina to see the difference that results in those countries where exploration is stimulated by incentive rather than placed in the hands of government agencies.

In the United States there is a growing tendency to penalize those who are fortunate enough to reap the rewards of taking the exorbitant risks of mineral exploration. Attention is focused upon producers who have been successful and receive benefits arising from the depletion provisions of the tax laws and the permissibility to write off exploration costs against other income. Despite the fact that these provisions were made for the very purpose of stimulating exploration, they are now called loopholes in the law. Furthermore, there is a growing body of opinion, currently exemplified by wilderness legislation pending before the Congress, that endorses the setting aside of huge tracts of possible mineralized regions as primitive areas. I am as anxious as any man to see our nation avoid the waste and destruction of our few remaining wilderness areas, and I do not propose to enter into this controversy beyond saying that nostalgia and sentiment cannot in the long run govern our decisions. The pressure of population, defense requirements, and the normal demands of industry are, I fear, incontrovertible facts.

It is interesting to speculate upon where the nation's industry would be today had we not stimulated the production of petroleum to satisfy the fuel requirements of growing fleets of automobiles and provided for the increased industrial uses of petroleum and its by-products. Without the incentives for taking such risks, we definitely would not have discovered the deposits that are supporting our present economy.

It is certain that a massive application of external economies is equally necessary to encourage further activities on the part of those who are not now in the mining business. The internal economics of the mining business, however, provide some incentive for the continuation of a well-directed exploration program because those companies that are now in the business must, in their own defense, continue exploration activities in order to maintain their reserves for the future. If long-range requirements are to be met, they must be encouraged to go beyond their immediately foreseeable needs.

One of the main drawbacks to the effective application of scientific research to exploration lies in the infancy of the basic science itself. There has not been the stimulus of necessity back of minerals research, a factor that has forced the advance of basic science in medicine, metallurgy, chemistry, and physics. Our knowledge of the fundamentals governing the origin of ore deposits is largely lacking, and an enormous amount of fundamental research is necessary to provide the applied exploration scientists with a reservoir of ideas on which to work. Several attempts have been made in recent years to channel the flow of funds into fundamental geological sciences. The National Science Foundation has recognized the need for this and has attempted some activities in this direction, but a great deal more scientific investigation into the origin of the concentration of ore minerals on the earth's surface must be undertaken before long-range scientific exploration can become sufficiently economic to satisfy future industrial requirements.

Usable resources are governed in the end by the actions of men and not by geochemical processes. The future of our industrial society will be determined by what we do to encourage research, through exploration and other means, into new minerals technology. Nothing will be accomplished through increased taxation or the setting aside of vast areas closed to prospecting for perpetuity.

Unfortunately, the minerals producer is finding it increasingly difficult to undertake new ventures, and before a final decision is reached in reviewing problems of conservation as applied to exploration, we must decide whether we really desire or need resources at a lower cost per unit for the future.

PAUL W. MCGANN

Economics of Mineral Exploration

THE ECONOMIC implications of mineral exploration constitute in many ways a neglected subject. This is partly because of a "mystique" surrounding the subject of exploration; its history is full of illogical, dramatic lucky strikes and of many bitter disappointments, both of which inhibit temperate discussion. Another reason is that economists have not been particularly attracted to the study of exploration—nearly all economic subjects are at best framed in an exasperating content of uncertainty, and the additional uncertainties of mineral exploration may seem just too much for all but a few economists.

PAUL W. McGANN, chief economist of the United States Bureau of Mines, has been engaged in minerals economics analysis for nearly ten years following a previous decade in industrial economics fields. He was chief economist for the Defense Minerals Administration, and a consultant to the Paley Commission. Previously he worked on national strategic problems as a member of the Operations Evaluation Group of the Massachusetts Institute of Technology attached to the Chief of Naval Operations. He taught economics at the University of California at Berkeley, and at American University he was chairman of the Statistics Department. During the war he was aerologist on the antisubmarine aircraft carrier U.S.S. Croatan. He worked in the Office of Price Administration national office as head of the Machinery Research Section and head of the Retail Building Material Section. His graduate work in economics was at the Universities of Chicago and Minnesota, and he received his A.B. at Brown University. He was born in 1917 at Newburyport, Massachusetts.

The economic implications of exploration may be classed under two headings. Exploration activity responds to price, and exploration results change the relationship between price and mine production from what it would have otherwise been.

As technology progresses and higher grade deposits are depleted, lower grade deposits become economic to work. The frequency of occurrence of deposits is such that there are very few high-grade deposits; the lower the grade (in terms of content of mineral sought) the more deposits there are. Thus as we go to lower grades the amounts of potentially usable material usually become larger. Ample concentratable amounts of iron ore are now available for as long into the future as can be foreseen—at slightly higher prices (in constant dollars). Aluminum raw materials are almost at that point as a result of United States Bureau of Mines research, but copper, lead, and zinc may not reach this status for many years. Thus, exploration activity may become relatively less important over time even though it will always be valuable in indicating the best ores to use out of the knowable, available low-grade ores.

We shall briefly consider the economic implications of exploration under five general headings: how to measure economic effects of exploration, available economic data on exploration activity, subsidies, projections, and sufficiency of exploration.

Measuring economic effects. First, what is the meaning of "supply of" a mineral? Supply of any commodity can be efficiently described by an equation relating amount supplied per time period to other quantifiable variables such as "real" price of the mineral commodity (actual price divided by some price index, to allow for fluctuations in the general price level), prices and amounts available of inputs to the mineral producing process in question, and optimal combinations of inputs and outputs. This mathematical relationship can be usefully condensed for many problems to a simple equation relating rate of production of the mineral to its "real" price.

This simplified procedure and its more complicated, larger equation both suffer from the fact that unmeasured elements

are also very important. The three most important of these
are depletion of deposits, the state of knowledge of mineral
deposits, and the state of knowledge of producing processes.
It is feasible to take account of the net effects of all of these
nonquantitative variables together by means of "shifts" over
time of the supply equation (whether the large, complicated
equation or a briefer, simplified equation), that is, by means
of changes in the constants of the equations. The development
of a mineral "shortage" now can best be described as a shift
of the supply function under which less is produced at given
real prices than before.

Unfortunately, adequate statistics of the effects of depletion
and technical progress are not available, although they could
be obtained if firms were to undertake more detailed reporting
of output by deposit, classified by year of "discovery." If mineral industries become seriously interested in the effects of
exploration and technological progress, and if the use of automatic digital computers becomes more widespread, a program
of estimating effects of technological progress could be insti-
tuted. Until this indefinite future we must largely resign ourselves to observing, pondering, and guessing about combined
effects of depletion, exploration, and technological progress
on supply functions. There are several interesting cases of
these effects that invite preliminary discussion.

The case of crude petroleum domestic output in the United
States has been one of a supply curve (between output and real
price) shifting steadily outward to greater production at given
prices until 1945; since then the supply curve seems to be fixed
in position with output increasing about 10 per cent for every
10 per cent increase in real price. For other major producing
areas (Venezuela, Canada, the Middle East, and Indonesia)
the supply curve still appears to be shifting outward. Bitu-
minous coal and anthracite's supply curves have been shifting
inward, mostly because of rising labor costs, but depletion has
been significant for anthracite. Both the United States and foreign Free World iron ore appear to have stable supply curves,
with the output of each changing 10 per cent for every 10

per cent change in deflated price, except for three unusual war years and three bottom-of-depression years. The curve for domestic mining of copper shifted outward until 1918, after which it has stabilized with an 8 per cent average change in output in response to a 10 per cent change in real price (except for four deep depression years). The foreign Free World copper supply curve has been stable since 1931, but output has changed 15 per cent in response to 10 per cent changes in price. The supply curve of domestic lead mining shifted outward until 1914, was stable until 1930, shifted unsteadily inward until 1946, and has been relatively stable since then (in the 1901-9 location) with output changing 5 per cent with a 10 per cent change in real price.

Data. There are almost no economic data on domestic exploration inputs, except for the petroleum industry and such data as exist are poor. The main source for nonpetroleum exploration data is material appearing in published company financial reports which is not comparable among firms.

The Canadians do this sort of thing much better than we do and apparently consider economic aspects of exploration more seriously, as a source of health for their mineral industry. Statistics of nonpetroleum exploration footage drilled in Canada are available annually for twenty years. Diamond core drilling is not all of exploration activity, of course, but, as in this country, it constitutes a fairly large share of total exploration expenditure. Drilling lends itself rather well to physical measurement, and it is generally easy to separate in companies' bookkeeping and engineering records because so much of it is on a contract basis.

Congress has not yet required the Internal Revenue Service to collect the exploration data which mining companies must prepare to account properly for their income tax deductions. If such information were collected, it could be used to assess the billions of dollars of indirect subsidies granted over many years through special tax treatment of exploration expenditure. One could then learn the extent to which indirectly subsidized exploration effort by industry compares with the direct government efforts of the United States Geological Survey, Bureau of Mines, and Office of Mineral Exploration. It may be that Congress now annually appropriates as much money for governmental exploration of nonpetroleum minerals as all of industry spends—but no one knows. If fiscal considerations have not stimulated Congress to require some agency to collect exploration data in the past, it is still possible that the keen interest Congress has demonstrated in the welfare of the mining industry through the hearings and bills sponsored through the Committees of the two Houses, will lead eventually to such statistical fruition.

Subsidies. One outcome of exploration and depletion trends is the nation's shift from the position of mineral exporter to that of mineral importer, a change experienced by western European nations a century ago. This economic implication of the spread of exploration activities gives rise to chronic political concern and outcry and attempts to persuade Congress both to raise barriers against imports from higher grade foreign deposits and to increase subsidies for uneconomic mineral output, either directly or covertly, through purchase programs, loans, special income tax allowances, and other schemes. This persuasion is based on two basic arguments which provide the core for much intellectual confusion. The first argument is that although the nation as a whole may benefit economically from the shift from domestic to cheaper foreign minerals, the burden of the cost of adjustment lies mostly (and unjustly) on the displaced domestic mineral industries and dependent communities. The second argument is that for reasons of defense supplies of these minerals should be mostly domestic.

There is some substance to these claims, but strong rebuttals can be made to both. The answer to the contention that the burden of adjustment cost falls unfairly is that such apparent inequity is inherent in the free market system. It is true, however, that our society can tolerate temporary, declining levels of assistance to extremely suffering businessmen, unemployed, and depressed communities who could not be expected to provide adequate self-protection against unusually rapid

change. If such a compromise is made, even the reduced substance of the first claim for other assistance to domestic minerals is removed.

The second argument is countered by the criterion that
provision for defense supplies of minerals should be achieved
at minimum cost. In recent years the government has come to
accept this view. Such acceptance has been somewhat vague so
that the public generally apparently has still to become aware
of it. This rule will generally indicate that a combination of
measures be used to achieve a minimum cost combination of
augmentation of the desired level of defense supplies. This
combination could include two or more programs of stock-
piling, maintenance of standby facilities, indirect tariff subsidy
of domestic industry, subsidized loans, or of other direct subsidy programs. Thus, within the context of "least cost," for
example, some minor tariffs may be justifiable, but it is only
in this manner that the defense argument for tariffs makes
sense. The feasibility of making such least-cost-combination
estimates has been demonstrated in governmental planning
and program assessment, but there is as yet no audible public
interest in such calculations.

Projections. In view of what has happened in the past to
supplies of minerals as a result of the joint effects of exploration and technical progress, it is interesting to speculate on
what will happen in the future. If the same past trends are
assumed to continue, it is simple to "project" (if not "forecast")
what will happen to future supply functions. Unfortunately,
no one can say for sure that past trends will continue, but they
may serve as first approximations.

The combination of the supply curves (shifting over time)
plus the projections of future demand, when solved together
simultaneously, provide us with a basis for estimating future
real price, that is, the price in constant dollars. From such
estimates of future price and the historical relationships between price and exploration inputs (such as petroleum drilling)
we may make some useful judgments of the future likelihood
of shortages (as defined above in terms of shifting supply) and

the implied economic role of exploration. Net effects of trends of depletion, exploration, and technological progress on future supply may be taken account of by using historical trends of shifts in supply curves.

This simple analytic procedure may be illustrated with the following two figures. Figure 1 shows the relationship between consumption and Gross National Product (in constant dollars)

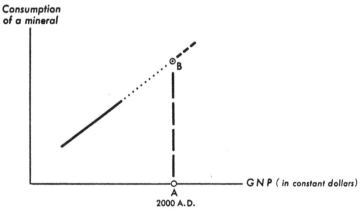

Figure 1.

for the United States. We can read off the projected consumption (B) associated with the projected GNP for the year 2000 (A). Figure 2 shows the projected shift in the demand curve from 1957 to 2000 at the 1957 price (from point C to point D). The new demand curve for the year 2000 intersects the supply curve at point E, reflecting relative declines in consumption at higher prices—at the 2000 level of GNP—rather than the consumption (B on Figure 1) associated with GNP of 2000 assuming constant price. Thus, price does not rise as much by 2000 as it would have (point F on Figure 2) if a larger quantity would be demanded, unmodified by effects of a price rise.

It is interesting to speculate further on what world demands for minerals will be in the year 2000 in order to judge the magnitude of this challenge for exploration. Such a world-wide look at the future is necessary because almost all minerals are influenced markedly by world markets rather than by intranational or local markets only. Unfortunately, no one

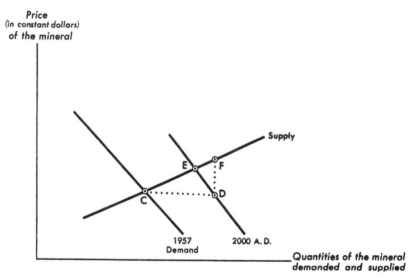

Figure 2.

(not even Colin Clark) has been brave enough to publish Gross National Product of the world for any one year, let alone a series of years from which a simple extrapolation might be made—as is customary for projections of the United States GNP (in constant dollars) based on the historical 3 per cent apparent average annual rate of increase realized over the past ninety years. Nevertheless, some orders of magnitude can be discussed. The author computes that a physical index of world mineral consumption in 2000 would be three times that of 1957.

Resulting sample percentages of estimated increases of important minerals prices in 2000 are listed below to illustrate the pressures that prices may exert upon mineral producers to expand output and stimulate exploration and research.

	Comparative price increases (in per cent of constant dollar prices)	
	1912–14—1957	*1957–2000*
Crude petroleum, U.S. Gulf, 32° gravity	20	25
Bituminous coal, average mine value, U.S.A.	45	45
Iron ore, Mesabi, non-Bessemer, lower Lake ports	45	50
Copper, f.o.b. refinery, U.S.A., electrolytic .	−25	50
Aluminum, primary ingot, U.S.A.	−45	40
Phosphate rock, land pebble, dried bulk, carlots, mines, U.S.A.	−35	10

Even if real prices of minerals in respect to prices of other goods and services were to double or triple by 2000, such an increase would have almost no effect on world levels of income. This is so because the mining and concentrating of minerals now consist of no more than 2 per cent of the current national income of the world. If this per cent doubled, the level of world income would be 2 per cent lower at that future date, but this per cent is less than the annual average increase in real income. Thus, fifty years of minerals depletion could be made up through one year of normal economic growth. The real concern for the future is economic growth, not running out of minerals—at least for the next hundred years.

Sufficiency. Will exploration be sufficient to enable the standards of living in the world to rise faster than population and to meet the fantastic needs a century hence? Many people assert that technology and science will inevitably solve the problem of perverse and persistent low levels of income. They think that some wonderful new developments can expand productivities in the underdeveloped nations constituting two-thirds of the world's current population, so that incomes can rise fast enough to enable them to break out of their vicious circle of high birth rates and low incomes. It is extremely unlikely that any investment in science can achieve this end. This is so because heavy investment in capital goods is necessary to apply scientific production aids. It is the absence of enough investment that is the problem. No remarkable scientific discovery nor mineral exploration will achieve the desired results, whether it be desalinization of water, harnessing of solar energy or nuclear energy, or hybrid strains of plants and animals. On the other hand, no baffling scientific mysteries need be solved to achieve the desired economic ends. Science can only make the solution easier. In blunt terms, it must be stated that economists can clearly say what the solution is to chronic low incomes—adequate investment.

All that remains is to persuade the underdeveloped nations to adopt this advice adequately by saving, by facilitating investment of foreign funds, and by restricting their population growth so that available investment need not be diverted from raising standards of living to the provision of the old, low standards for increased numbers of people.

The economist's advice about investment for underdeveloped nations must seem exasperatingly smug and superior, coming from a nation which has for 300 years been free of the vicious circle and capable of adequate investment. Do we have to buy these nations off so that they will accept good advice, by sweetening the remedy pill with additional grants of billions of dollars? No economist knows the detailed answer to this question of political strategy, but there is a far sweeter coating in the promise that any nation which keeps its population stable and saves over 20 per cent of its income (so that it can increase its national income 6 per cent per year) can increase its per capita income tenfold in less than fifty years.

The complaint that underdeveloped nations cannot save one-fourth of their incomes is unfounded. Economies save at all levels of per capita income. It will hurt, but they already save nearly 10 per cent, and there is always another economic group within such a country, that the rest of the nation can look down upon, which has an income 15 per cent less per capita than theirs. The very lowest group—which constitutes less than 2 per cent of the world total is a very small exception and could be helped up to a saving level by investment loans from richer nations.

Everyone who discusses the distant future of natural resources must come to grips with the pessimistic predictions of Sir Charles Galton Darwin, who forecasts that population pressure will lead to such declines of income levels per capita and to such widespread war that human culture will revert permanently to a Dark Ages condition. This is the ultimately dramatic statement of the problem of future needs for natural resources. It is safe to say that inadequate exploration would not be a primary cause of such a situation. What really could bring about a Galton-Darwinian future within the next 100 years is inadequate investment in the underdeveloped countries, resulting from the pursuit of unsound economic and social policies by these nations.

Now, let us briefly look 1,000 years into the future rather than just 100 years. At that date no one would say that we can depend heavily upon coal and petroleum for the energy base of our world economy. Unless "breeding" of nuclear materials becomes economic, we would also probably be running out of economic ores of uranium and thorium, regardless of any practical exploration efforts. This is the future period for which concern with running out of minerals becomes very serious. This is the "conservation nightmare"—that our civilization would follow that of the Mayas into oblivion, having depleted its basic natural resources that are within the scope of its technological knowledge. At present the most widely recognized materialistic hope is simply to trust in science. There is, however, one very important alternative to this Mammon-like fountainhead, namely, an ecological balance at a high per capita income level. This alternative is a practical solution, not just an act of faith.

This ecological balance means that solar energy capture could be made adequate for a permanent supply of energy of a world population several times the present at a per capita income level several times that of the United States today. Based upon present basic technology and very heavy investment, most agricultural land would be devoted to fuel plant production and much additional nonagricultural space to solar cells of various sorts. Energy costs would constitute a high fraction of Gross National Product, and real energy prices would be many times present levels, but there still could remain enough income for much higher standards of living for many more people.

Conclusions. Going to lower grade ores requires more energy per pound of refined mineral, despite technical improvements in process. Thus, the availability of nonfuel minerals ultimately becomes a problem of availability of useful energy. Exploration will be very important for finding the better grade fuels and other minerals during the next 100 years. Over the next 1,000 years and beyond, the mineral exploitation outlook is essentially similar, but far more intricate (and probably highly capitalized) processes will be used to refine huge amounts of low-grade materials. Exploration may be even more important in this era to show which "deposits" to use with minimum cost and disruption to other uses of the earth.

iv

CHEMICAL TECHNOLOGY

Earl P. Stevenson:
PAST GAINS AND FUTURE PROMISE

Frederick T. Moore:
UNLOCKING NEW RESOURCES

Richard L. Meier:
THE WORLD-WIDE PROSPECT

EARL P. STEVENSON

Past Gains and Future Promise

CHEMICAL TECHNOLOGY penetrates all areas of our industrial economy. Its products have a role in satisfying the six basic needs of mankind—food, clothing, shelter, transportation and communication, medication, and tools, machinery and equipment with which to work. They serve the seventy-two basic industrial groups as enumerated by the United States Chamber of Commerce. Their markets embrace our entire economic structure.

Knowledge derived through the science of chemistry is not

EARL P. STEVENSON is chairman of the board of Arthur D. Little, Inc. He has been with that long-established industrial research organization since 1919; vice president 1922-35, president 1935-56, and chairman of the board since 1956. From 1916 to 1918 he was a chemistry instructor at Massachusetts Institute of Technology. He served with the Chemical Warfare Service, United States Army, in 1918, and in World War II he was with the National Defense Research Committee from 1941 to 1945, most of the time as chief of the Chemical Engineering Division. Mr. Stevenson was awarded the Medal for Merit. He is active in many professional societies, and in civic affairs in the Boston and New England area. He was born in Logansport, Indiana, in 1893, and received his B.S. from Wesleyan University and M.S. from the Massachusetts Institute of Technology, and from Wesleyan where he is now chairman of the Board of Trustees.

restricted to a narrowly defined chemical industry but permeates most other technologies. We often speak today of the chemical and allied industries to include particularly the petroleum and pharmaceutical industries. But chemical technology extends beyond the orbits of these satellites. In this metaphor, agriculture and the extractive processes come into easy view. Its practitioners have invaded what was for the scientifically trained person outer space—the security markets. Actually it is from this vantage point—the portfolio of a mutual fund dealing exclusively in the securities of companies associated with chemical technology—that we can obtain a most comprehensive and inclusive view of our subject. The sixty-two companies chosen by investment criteria from a larger eligible list of companies embrace the so-called general chemical companies, drugs, specialties, oil and gas, glass, pulp and paper, corn products, metals and rubber. This list bears witness to the key position of chemical technology in our national economy.

The second characteristic which should be noted is that no other major technology is so close to a basic science. It is the only technology bearing the name of a scientific discipline. Hence, any appraisal of the future of chemical technology explores the expanding frontiers of a science—a science that in turn is involved with other scientific disciplines.

With the dependence of these industries on creative research, a measure of growth potential is the continuing attraction of chemistry for those who elect a career in science. Historically, the majority of graduate degrees in science have been in the field of chemistry, and the majority of those receiving such degrees have entered industry. Men trained in the science and technology of chemistry permeate the whole fabric of the industries they serve. The science of chemistry opens to chemical technology an endless frontier.

The third characteristic, which I want to develop in greater detail, is that chemical technology is a powerful extender of the resource base. Products that are quite different in composition can be derived from different starting materials or can be synthesized by different processes from the same starting materials. Take the gasolines that we use in our cars. Today these are truly synthetic fuels. Straight-run gasoline, as we speak of the natural product, derived from straight distillation of crude as it comes from the ground, would not be usable in the present high-compression engines and would not respond to the energy requirements that we demand—easy starting, quick and smooth acceleration, freedom from knock, etc. Aviation gasoline depends even more upon chemical technology for performance and without its contribution, modern aviation would still be in the "Model T" stage of development. Here it is significant to note also that fuels having the same characteristics in use can be derived from a wide variety of crudes differing in their molecular composition.

Identical end products can be produced by different processes starting with the same raw materials or, more importantly, from entirely different raw material bases. Many important petrochemicals, for example, can be made from either ethylene or acetylene, the former derived usually from the cracking of hydrocarbon gases or as a by-product from the cracking of crude oil to produce various distillates, notably gasoline; acetylene is usually derived today from calcium carbide for which the requirements are limestone, coal, and electrical energy. In the future, acetylene derived from the cracking of hydrocarbon gases is expected to be an increasing factor in this area of chemical technology.

Possibly the most dramatic illustration that I could cite of this attribute—a degree of independence so far as raw materials or starting points are concerned—is the case of nylon. Today there are two competitive routes, one starting with benzene, which in turn is derivable from different sources. The important intermediate from which the ultimate product, nylon, is produced is hexamethylene diamine. The benzene route leads through benzene, phenol, cyclohexanol, cyclohexanone, adipic acid, adiponitrile, and thence to hexamethylene diamine. The alternate route starts with furfural which can be produced from such lowly products as corn cobs and oat hulls. Other agricultural by-products can also yield this material.

Starting with furfural, the route to hexamethylene diamine leads through furane tetrahydrofurane to dichlorobutene to adiponitrile and thence to hexamethylene diamine. The diverse chemical processes involved in the ultimate production of salt from which nylon is produced have their starting points in the coal tar industry, the petroleum industry and in agricultural by-products as diverse as oat hulls, corn cobs and bagasse. These characteristics were summarized in *Resources for Freedom*, the report of the President's Materials Policy Commission (1952): "The chemical industry differs from other industries chiefly in the fact that it has tremendous flexibility, both in the raw materials it utilizes and in its processing techniques. It has a unique facility for processing abundant raw materials not only into products suitable as substitutes for other materials, but also into new materials having new properties and superior to anything previously known . . . . Chemical technology works with what it can get: sand, salt, soda, brines, air, water, sulfur, phosphate rock, limestone, cellulose, coal, petroleum, lignite, natural gas, molasses, starch, and even corn cobs and oat hulls."

It is on the basis of these characteristics that I now venture the proposition that the outlook for chemical technology is the hope of man's survival. We can look for no relief from the pressure of population and rising standards of living through any system of denials. Looking ahead two decades and assuming no catastrophic events, a major concern is the ability of the country to grow and prosper in terms of traditional patterns. To achieve such a goal we must consume more, and this worries the conservationist; also want more, and this worries the idealist who sees ahead only an increasing materialism. Here it is interesting to note that the industry of want-creating—advertising—is in good balance with technical research, the annual expenditures in both areas amounting to about $10 billion.

These aspirations worry the demographer. By 1975 there will be between 55 and 75 million more people in the United States than in 1950. Broader limits are possible with an upper limit estimate of 243 million by 1975. By 2050, present trends extrapolate to a population of about a billion. Looking ahead then to the era of our children's grandsons, the national consumption on the scale of present social values is truly astronomical and numbers appear ridiculous.

The continuing fertility of technological research appears to be the only answer except as we choose another course in "the pursuit of happiness," as our historic goal is defined. Here I tread lightly as there is quicksand underfoot, but ultimately the only way out may be the wide acceptance of the idealist's goal of a better life through nonmaterial pleasures. Critics of our culture tend to overlook the contributions of science and technology in making available to large numbers—not a small elite group—the major resources of the great literature and arts of the past and present. The growth of scientific intelligence has helped to bring about not only improvements in the material circumstances of life but also an enhancement in its quality. The magnetic tape and the record disc owe much to chemical technology.

Technology has contributed in other material ways to nonmaterial pleasures, and chemical technology is a major participant. From the aspirations of the early thirties—a chicken in every pot—we now have light, weatherproof sporting equipment in every car trunk, and in the trailer behind an easily portable skiff or boat for fishing or other aquatic pleasures. Waterproof, synthetic adhesives have revolutionized the wood veneer industry, and practically all small boat construction today takes advantage of these new materials or use alternatively resin-reinforced glass fiber construction. The glass fiber rod has brought the delights of fly casting to millions. Fortunately, fishing is pleasurable to a degree without tangible results. Pesticides, in controlling pests, may conserve the forests, but the poisoned larvae may kill the young fry. The fisherman-chemist now turns hopefully to the ichthyologist for the solution of this problem.

With these growth opportunities in mind, we can attempt an appraisal of the outlook for chemical technology. Universality, a close working alliance and dependence upon a basic science, and substantial independence in terms of raw materials, afford the foundation for the continuing and healthy growth of chemical technology.

To start this phase of my discussion, I might ask a double question. What are the forces underlying technological growth? What are the sinews of technology and chemical technology in particular?

The explosive growth in population, increasing wants, and changing patterns of living create the demands for the products of chemical technology. An appraisal of this opportunity must, however, reckon with some material restrictions. Here a major item is energy. The basic chemical industry, from which chemical technology largely derives, is a large consumer of energy. And for many of its operations, if it is to exercise fully its independence of any one raw material base, power must be cheap. Major enterprises based on chemical technology are located where power is relatively cheap. Historically, the modern chemical industry in this country had its beginning with the generation of electric power at Niagara and that area saw the beginning of the electrochemical industry in this country.

A review of the major trends of the present-day chemical industry will serve to develop further the theme of "opportunity unlimited." A very sketchy outline of these major areas will have to suffice and several will only be identified by name. Here I will have to turn from generalities to particulars and identify as significant growth areas high energy chemicals, petrochemicals, polymers and new metals and organo-metallics. Drugs, pharmaceuticals, fertilizers, pesticides, will only be mentioned, though they, too, are pertinent to this discussion.

More powerful rocket-propellant fuels will be needed to meet the demands for ever-increasing power in small packages. Within the next decade, chemical propellants will probably have reached their peak of development and limits of capabilities. From that point on, more powerful systems such as nuclear or ion propulsion systems will be needed to meet that demand for ever-increasing power. Construction is now well along on two major new plants to produce alkyl borane compounds for use in military aircraft. The higher heating value of boron fuel, 25,000 Btu's per pound versus 18,000 Btu's per pound for conventional jet fuel, means an estimated 40 per cent increase in thrust. Other nonconventional fuels are also based on chemical technology. These include liquids to be used in bi-propellant systems, solid propellants and those, such as methyl nitrate, which have self-contained reducing and oxidizing radicals within the molecules themselves.

As in rocketry, national defense needs supply the main driving force behind rapid expansion in other areas of the established chemical industry. Notable among these is the interest in polymers and the underlying science of big molecules. While the urgency may largely derive from military areas, the incentive is shared with civilian interests. Over the past decade, consumption of plastic materials has grown from approximately 1.2 billion pounds to more than 4.2 billion pounds. In 1959 we have arrived at the beginning of an age where plastic materials are available that have not only the ease of fabrication associated with the earlier plastics, but also the range of physical properties that makes them competitive with metals, wood, glass, and other materials. These new plastics give the design engineer and the consuming public products which have the properties of noninflammability, high strength, high temperature resistance, and good electrical properties. Most important of all, these materials can be fabricated on standard equipment with relative ease. The new products do not so much challenge the existing markets for their predecessors—polyvinyl chlorides, phenolics, polystyrene and methacrylates—as they challenge materials used by industry which have not yet faced competition from any plastic material—steel, brass and bronze, aluminum, and glass. Even a minor invasion of these markets will have a profound effect upon the growth of the market for plastics in the next decade.

We are now entering into a new stage of polymer modification—custom tailoring. Modifications of this order are accomplished by rearrangement of the molecule prior to its polymerization, or during its polymerization. Polyethylene, for example, has achieved its present position through basic modifications in its molecular structure. Polyethylene plastics are available in a variety of types depending upon the end-use properties desired. Indeed, the whole area of polyolefines faces a brighter future because of our growing and better understanding of the polymerization process. A recent instance of this progress is the appearance on the market of crystalline polypropylenes which possess a combination of new and desirable properties. They have physical properties directly attributable to their molecular structure and which fundamentally differ from those of previously known thermoplastics—notably in their resistance to heat, with melting ranges in excess of 300°F. Mechanical properties are also superior — tensile strength, rigidity, hardness, impact strength. With practically no water absorption, the electrical properties remain virtually unaffected by humidity.

Other new plastic materials that are creating considerable interest include a polymerized formaldehyde and a chlorinated polyether. It is certain that the number of products available to the plastics fabricator ten years from now will be bewildering.

Finally, in the area of polymers, I should at least mention progress in the field of synthetic fibers. During the postwar period in this country, a social revolution has affected the ways in which we live, eat, dress and in general provide for ourselves. The basic changes in the way we live and work have created an interest in and a demand for entirely new types of clothing.

Here we are witnessing the kind of competition which is so characteristic of industries based on chemistry. Fibers designed for the same use, produced from different materials, having somewhat different characteristics, and being promoted by different companies, are in keen competition for markets basically influenced by changing habits yet at the same time subject to the vagaries of fashion. The changes in the ways we live and work have created interest in a demand for entirely different types of clothing. Casual clothes are increasingly the order of the day. The demand for more formal attire, such as dresses and suits, has actually declined despite the population increases in the period, while the weight of our clothing has also decreased. Summer suitings used to average 10 ounces per yard and now average near 7. With well-heated homes and work areas, we no longer need heavy clothing. Through the development of new fibers, the textile industry has been able to offer the consumer completely new types of fabrics as well as innovations in garment design.

The markets pioneered by nylon were first invaded by the acrylic fibers in 1946. And then came the first of the polyesters (dacron), being followed in 1959 by several additional materials, each with a different chemical history. Therefore, we see ahead a period of intense competition between various fibers and the many producers. In the years ahead, we can anticipate many more new synthetic fibers competing with the ones holding the spotlight today.

Much of the demand here follows shifts in the social structure. With 11.5 million women now employed, fabrics for the family must be easy to handle and care for—long lasting, wrinkle-resistant and often wash-wear. The synthetic fibers offer most of the answers, but the acceptability of wool and cotton has been increased, both by chemical treatment of the older fibers, and through blending with synthetics.

Molecular engineering is not confined to carbon chemistry. Silicone-type materials are examples of polymers developed for high temperature applications. At the time of their development, it was possible to predict from theoretical considerations that the introduction of a silicon atom into a polymer backbone would provide good thermal stability.

The effect of molecular structure on physical and chemical properties is universally important. Closely allied are the physiological or biochemical properties. Synthesis of new drugs may be more important to mankind than any other application of molecular design. The present research effort in proteins and nucleic acids, which seeks to shed some light on the mechanism of the life processes, may be important beyond comprehension.

Passing mention should at least be made of glass, man's old-est thermoplastic material. For centuries the peculiar characteristics of glass have fascinated man and led to ingenious uses, especially in the arts and in optics. Yet through hundreds of years the glass industry remained essentially static. Nearly all of the advances in the basic knowledge and useful applications of glass have occurred in the last few dozen years. The introduction of borosilicate glasses, with their resistance to thermal shock, brought a degree of efficiency and safety to laboratories of all kinds that can never be measured in dollars. Perhaps the most dramatic application is that resulting in the creation of the glass fiber industry.

Glass manufacture is an ancient art, but through new knowledge and more basic understanding of the chemistry of glass, new forms of glass are either now available or in the making. We have put this class of materials to work in ways never dreamed of a few decades ago. Since glass is stronger than steel, the ultimate challenge in glass technology is to tailor its molecular structure so that it can withstand fracture; then it may replace steel as a structural material in many applications. In the meantime, a major company is exploring commercially the uses for a new form of glass which will challenge metal in a number of significant applications—ball bearings, for example.

During the past few years, emphasis has been placed upon the creation of basic facilities for the production of hydrocarbons and their primary derivatives which require further processing before going into such markets as plastics, fibers, rubber and detergents. Probably the most striking has been the dramatic increase in capacity for producing ethylene and the building of pipelines to distribute this basic raw material to different manufacturing plants for a variety of purposes—notably the production of ethylene oxide and polyethylene. Ethylene oxide, first used in the manufacture of ethylene glycol (antifreeze), is finding increasing use as a building block from which to construct more complex molecules. Wetting agents and detergents are among the other outlets for this basic material.

Isoprene now gives promise of becoming a petrochemical of

major commercial stature. It is now a by-product of gas oil cracking but processes will be developed for synthesizing it from other hydrocarbons. The goal is a product that will have the same properties as natural rubber and compete with it in cost. In the meantime, the nation continues to depend upon synthetic rubbers now available in improved forms, following their wartime development. In like manner, new synthetic lubricants are making their appearance to meet the extreme operating conditions encountered in jet engines, rockets and missiles.

Another most interesting prospect for future development in the petrochemical industry is the emergence of hydrocarbon acetylene as a building block for organic chemicals. Already available as a pipeline product at major producing points for calcium carbide such as Niagara Falls, Ashtabula, Ohio, and Calvert City, Kentucky, there is increasing interest in the production of acetylene as a by-product of hydrocarbon cracking. Such projects place acetylene at the disposal of the chemical industry at a relatively low cost in areas where carbide-derived acetylene has been relatively expensive. Another vista opens!

New metals afford another example of expanding areas for applying chemical technology through science. The iron and steel industry is not conventionally thought of as a chemical industry, but with the pending advent of direct reduction, this concept may change. Few areas of the world are logistically favored with resources of high-grade iron ores and coking coal. In the United States there is the further coincidence that where iron ore and coal meet under favorable transportation circumstances, there is also the center of population and the industrial demand for the final product—steel. Depletion of natural resources, however, would end this heyday. Even now, this is being anticipated by the development of methods for reducing iron ore and utilizing this in lower grades through natural gas or even the utilization of coal or fuel oil for the production of a synthetic gas. This process will make use of direct reduction and utilize the fluidizing technique for the handling of the solids. While probably of the most immediate interest to other areas of the world, it can be anticipated that new centers for the production of iron and steel will ultimately develop in the United States, probably along the Gulf Coast.

Within my lifetime, many new metals have become available for practical use—light alloys of aluminum and magnesium, alloy steels, and a host of others. Part of this progress can be attributed to the availability in commercial quantities of elements which until quite recently were chemical curiosities. The needs of nuclear engineering are responsible for several of the more recent recruits to industrial service: uranium, thorium, zirconium, beryllium. The transistor has introduced us to germanium and silicon. The transistor may in time be challenged in some of its applications by a new device which in turn requires a new set of elements—columbium and tantalum.

The space and missile age confronts us with many new and urgent material problems. We have a desperate need today for better materials for use at high temperatures. In turning to the more refractory metals, chromium, columbium, tantalum, molybdenum and tungsten, increasing demands will be made on chemical technology.

Organometallic chemistry is also opening up new markets and use for metals. The commercial beginning was in the drug field at the turn of the century when it was found that certain arsenic and mercury-based compounds were valuable drugs. Mercurochrome has long been a household word. Tetraethyl lead, introduced in 1922 to suppress engine knock, was the metal industry's first substantial customer for purposes other than fabrication; it now consumes around 200,000 short tons of lead annually. A new fuel additive, not yet ready for commercial use, is based on manganese and cyclopentadiene.

Organometallics incorporating aluminum, magnesium, lithium, or beryllium, may find military applications as fuels. Trimethyl-aluminum and triethyl-aluminum are highly flammable liquids, bursting into flame when exposed to air. Their admixture in jet fuels prevents "flame-out" and hence loss of power at high altitudes. Jet engines burn a petroleum fraction heavier than gasoline and under certain operating conditions at high altitudes and speeds, the flame may be extinguished. To prevent this, spontaneously combustible additives are blended with the hydrocarbon fuel.

The boron-based "chemical" fuels belong in this category. These have already been mentioned in another association. To maintain its defenses, the United States is continually seeking means of making aircraft and missiles fly faster and higher, travel greater distances, or carry heavier loads. One way to improve performance is through the use of fuels more powerful than hydrocarbons. A very substantial government program has now brought these new "high energy" fuels close to practical use. They are the most publicized development in organometallic chemistry.

There are many other relatively minor but intriguing classes of these compounds. Tetraphenyl tin is used as an anticorrosion additive and certain butyl tin salts are used as stabilizers in plastic formulations and as silicone-curing catalysts. Vinyl tin compounds promise to be useful in certain plastics to increase structural strength and heat resistance. Other organometallics are effective catalysts.

With this citation of a few areas of progress and promise, I should like to conclude with three considerations which will, in the end, determine how well and with what timeliness chemical technology will meet the demands of a rising population—how it will increase and not just maintain our standard of living—and contribute its full share to the problems of national security, both on the civilian and the military fronts. These reservations—for such they are—concern resources in people, monies, and knowledge.

Starting with basic research, progress is through applied research, development work, engineering, plant design, production planning, and finally marketing. Technology as I construe it, comprises this entire spectrum or sequence. This is peculiarly so in the case of chemical technology as the professionally trained chemist will be found in key positions all along this production line. The science of chemistry is the life blood of chemical technology.

At this point, a cliché may be permitted as we view with alarm the possible shortage of men possessing the necessary scientific or engineering training. Following the rise for two years of engineering enrollment, there was a definite decline in the fall of 1958 and this appears to be most marked in chemical engineering. Also, the newer technologies associated more intimately and directly with the so-called space and missile age are attracting young talent at the expense of the more staid enterprises. This, however, is only a special aspect of a much more general and alarming situation. Technological progress depends upon men, and increasingly upon those individuals who can be identified as scientist-engineers. In our society, many careers compete for the limited resource from which such individuals are recruited. Our competitors, the Soviets, have, on the other hand, geared their whole educational program to meet anticipated needs in technically trained manpower, and, in this area, offer the nation's leading career opportunities.

In our free enterprise and capitalistic system, industrial progress always depends upon the maintenance of a favorable economic climate. Industries and companies identified with chemical technology have stood high in the investor's favor. Will they continue to so fare? Plant expansions and new enterprises must depend on managerial confidence in near future trends. The manager is ultimately responsible for the dividends which the investor expects. The basic industry of . chemistry and those most closely allied with it form the hard core. The record of the companies comprising this core has been outstanding from the investor's standpoint and has received his support. With the overcapacity now prevailing, a slight cloud appears on the horizon. Fortunately for the country, however, the fiscal year is not the planning unit of time for the management of a technically based industry. During 1958, plant expansion in many critical areas went forward despite the threat that current earnings might not equal well-established dividend rates. Recently, chemical stocks have again shown encouraging strength.

In appraising the past records of a company and an industry, the investor cannot take continuing progress for granted. The past has taught him that among other factors, research capabilities are of the greatest importance in appraising the value of the common stock of a technically based industry. This is the message of the annual reports of these companies to their stockholders. As instances, a major pharmaceutical house reports that 60 per cent of its current sales is in products less than three years old, and 80 per cent in products less than six; a major chemical company reports that one-third of total sales and nearly one-half of net income derived from products and processes not available fifteen years ago. To maintain this accelerated pace, new knowledge and new understanding of chemical phenomena are prerequisites. While large expenditures in the name of research will not insure progress, the consequences of a decrease in financial support are predictable. Chemical technology is not immediately faced with this threat but can expect the support of an increasing research effort. In the next decade this country is expected to spend twice as much for research, development, and associated engineering activities as in the past decade—an increase from sixty to one hundred and twenty billion.

We can be sure that chemical technology will be the beneficiary of this increasing search for new knowledge and understanding at the frontiers of the science with which it is so intimately associated. In the meantime, we stand in awe before the capabilities of the living cell and the mysteries of photosynthesis. The lowly oyster has abilities beyond the most intricate man-operated chemical process. The historian will probably refer to our times as the pick and shovel age of chemical technology.

FREDERICK T. MOORE

Unlocking New Resources

MR. STEVENSON puts his finger on a key characteristic of the chemicals industry when in the opening sentence of his paper he speaks of chemical technology as "penetrating" all areas of our industrial economy. On this point there is no cold war. The Russians feel the same way about *their* economy.

Let me quote from Nikita Khrushchev and some other Soviet writers: "The chemical industry is one of the decisive branches of heavy industry. Chemical engineering is one of the basic directions of technical progress . . . [it] ensures the creation of unlimited possibilities for broadening the raw material base of many branches of industry." And finally: "Now one of the most important and urgent tasks is to secure in a short time the rapid development of the chemical industry

FREDERICK T. MOORE is head of the Economics Division, Washington Office, of The RAND Corporation. Prior to that he was with the Bureau of Mines and was concerned with the interindustry research program. He has held teaching positions at the universities of California and Illinois. In 1957 he was a staff member of the President's National Security Panel, and at various times has been consultant to Resources for the Future and the National Economic Development Bank of Brazil. His published works have been on the theory of economic growth, regional development, and interindustry analysis. He received his M.A. from the University of Wisconsin and his Ph.D. from the University of California.

and to create a strong industry for manufacturing polymeric substances." This quote is from an article titled "The Development of the Chemical Industry—One of the Most Important Economic and Political Tasks" (English translation from *Voprosy Ekonomiki,* December, 1958). It is interesting nowadays to compare our situation in any area to that of the Soviets. Otherwise there surely is no need to seek added support for the basic thesis of the importance and ubiquity of the chemical industry.

Mr. Stevenson has portrayed technical developments in a variety of chemical commodity areas. Given these, an economist inquires into the probable impacts of such developments on economic and related activity. This requires prediction of the future and interpretation of purely technical phenomena in terms of their economic and social content over a time period of, say, the next quarter century. Anything short of this makes it extremely difficult to identify persistent trends.

We often speak loosely of the "technical growth" of an industry and many times a commonsense meaning of the term will suffice. But in fact there are several different viewpoints for measuring growth of an industry.

First, we might look at growth of the chemical industry from the viewpoint of the final consumer. Here there is an overwhelming abundance of materials. Magazine articles and the press almost daily point up the variety of new products which are now or will soon be available to the consumer. Orlon, dacron, teflon, lexan and all the rest of the trade names identifying the synthetic fibers, plastics for toys, utensils, piping, and packaging, food preservatives and new plastics in construction add up to an impressive list. From the handle of his morning toothbrush to his nightly pajamas, a typical consumer is surrounded by the new product. Judged by this standard the growth and impact of the industry have been nothing short of spectacular. There are, however, two consequences which deserve passing mention.

Perhaps you will recall a British movie of a few years ago

called "The Man in the White Suit." In it a chemist succeeds in producing a cloth which appears to be perfect. It will not wrinkle nor wear out and it never needs cleaning. It thus represents the highest state of the art; but the economic consequences are far from perfect. In fact, by its very excellence the cloth leads to widespread technological unemployment in the textile industry. The point is this: we pay a price for the new products—a price measured in terms of technological disruptions in other industries. To an economist it is the price of growth and should be paid gladly, but the price is not always low, as we can see in the state of the textile industry in New England and in the various political schemes for bailing it out of its troubles.

Furthermore, it is apparent that in both synthetics and in pharmaceuticals there is a proliferation of trade names for the same material. In order to prevent market penetration by competitors, each firm finds it necessary to engage in extensive advertising expenditures. Now whenever a new product is put on the market some advertising is necessary in order to educate consumers, but the kind of advertising (particularly in pharmaceuticals) to which I refer is not of that kind. It is primarily to offset the advertising by other firms and consequently on the whole it represents net economic waste of resources.

Another way for an economist to measure or project the growth of the chemicals industry is to look at past rates of growth and to try to estimate future growth rates in some consistent fashion. We can observe, for example, that aggregate chemical output has increased by about 10 per cent per year in the past decade or so. Some individual commodities have had far greater increases than this and some less. However, compare this rate of growth with some others. Our population has been growing at approximately 1.1 per cent per year; GNP at something like 3 per cent per year and industrial production at a slightly higher rate. Thus, by comparison, the output of end-chemical products has been outstanding.

Projection of future growth for the industry is hazardous and may be misleading since there are numberless techniques for making such projections. Nevertheless, it may be instructive to compare several projections, all of which are based on relatively simple relationships. Four such extrapolations to 1980 are shown in the accompanying figure. Three of them are based on relationships between end-chemical output and some combination of GNP, time, or time squared. The purpose of

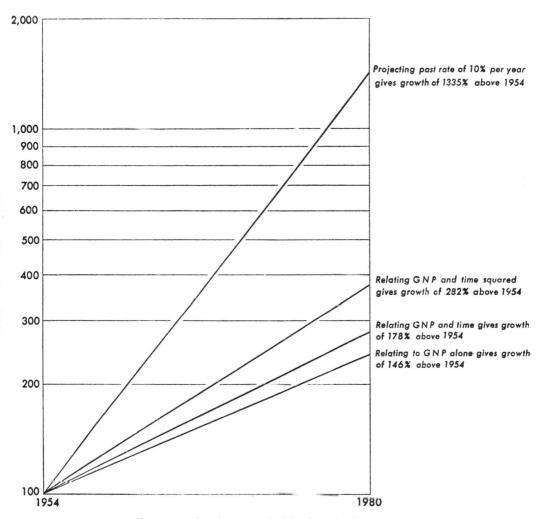

Four projections of 1980 chemical output.

such a variable as time squared is to try to cover, by an omni-
bus statistic, the effects of other factors which may have a great
impact on chemical growth. The fourth, and highest, projection is based purely on the growth rate of the recent past of
10 per cent per year. All four of the projections show chemicals
growing at a faster rate than that expected in GNP, and if I
were asked to guess as to the most probable value for the
growth of chemicals it would, I think, be at least twice as great
as that of GNP. But one should not view these projections
too seriously. They are useful primarily for illustrative purposes or as indicative of rough orders of magnitude.

A third way of viewing the development of the chemicals
industry is in terms of the rate of innovational change. As
contrasted to the sweep of human history which would be
measured by the life-span of a number of men, the history of
modern chemistry can almost be covered by the span of a
single man who might be alive today. The rate of innovational
change has been exponential. But it may be dangerous to
assume that it can continue to be so into the future. Perhaps
the investigations of molecular structure will reveal new materials which are as different from those we know today as
polyethylene is to the older materials of wood and metal. Or
perhaps there is a top limit to the rate of innovations in chemistry. If there is, then sometime soon—maybe in the next twenty-
five years—there will be a definite slowing up in this rate, and
we will instead be simply exploiting productively our past discoveries. We can observe, for example, that in other industries there were periods in their early growth when the rate of
innovational change was very high; examples can be found in
metal processing and fabricating. Later on there was a slowing
up in the development of new processes and materials. The
same may occur in chemicals. What I am suggesting is that
projection to the future of high initial growth rates may be
misleading. As chemical products are substituted for other
products such as wood and metals the possibilities of exploiting
further substitutions tend to narrow and the rate of growth
is retarded.

Let us now speak of some of the specific characteristics of the chemical industry which are of importance to the growth of the rest of the economy.

One statistic which indicates the potential growth of an industry is the amount of funds devoted to research and development. In general, the larger such amounts are, the greater is the flow of new products. The impact on consumers is immediate as we have noted. But the indirect effects on technology in other industries—the stimulation to the development of production which is ancillary to chemical products—may be equally as important to the nation as a whole.

Of all the major American industries the chemicals industry and allied products have spent the most from private funds for research and development. According to a National Science Foundation survey of industry, chemical companies spent on the order of $350 million in 1953 for these purposes. The motor vehicle industry may have spent more, but precise figures are lacking, and the range of its research was certainly smaller than for chemicals. Perhaps of even greater significance than these totals, which include a lot of development work spent on preliminary engineering of products, is the amount spent on basic research; that is, research to improve the state of the art. Here the chemicals industry far exceeded any other industry. In 1953 it spent approximately $38 million on basic research or twice that of the next industry, electrical and electronic equipment.

It is just because long-run growth depends on breakthroughs in the state of the art that expenditures on basic research are so important. But basic research is a painstaking and time-consuming process; its returns are uncertain. Thus when, after years of research, a product such as nylon is perfected it is proper that the research results should be protected by patents which are the reward our society bestows for creative effort of this kind. A patent is a right to a limited monopoly in the exploitation of an idea. It is expected that after a time the process which the economist Schumpeter called "creative destruction" will result in new products and patents which will supplant the old. That way lies progress.

But there is another side to patents. In the chemical—and in other—industries there have been times in the past when patents were the basis for cartel arrangements aimed at controlling the market in the interests of the members of the cartel. And sometimes patents combined with, say, the exigencies of government policy during national emergency have resulted in a new form of government-industry relationship. The development of the synthetic rubber industry during the war is an illustration of such a new relationship. The government paid for and owned the plants initially; industry operated them and made most of the critical policy decisions; finally the plants were sold to some of the largest firms. Cartels or the like are not inevitable, but on the other hand the chemical industry is not a model of a fully competitive industry. New types of market institutions representing closer ties between government and industry—frequently in the name of national security—may replace older forms of market organizations and if they do they will bear close watching. New public policies may well be required. Some observers of events see in current developments the emergence of "market syndicalism." Apropos of the synthetic rubber case, Senator Joseph C. O'Mahoney of Wyoming remarked in February, 1959, that the performance "fell far short of what the public interest and national security demanded." The next twenty-five years will probably see further developments in these directions.

Another aspect of technical development which is particularly important in the chemicals industry is the growth of automation. The improvements in computers and servomechanisms have made it possible to exercise control over extremely complex operations. Chemical technology especially lends itself to the application of these new techniques since most chemical processing is of a continuous flow with the gas, fluid, or solid being subjected to heat, pressure, the addition of catalysts, etc. at the several stages. Also the materials individually are homogeneous. Chemicals have never used very many employees per dollar of output. Roughly speaking, fifty to seventy-five persons per $1 million of output are all that have been required. And with the increase in the automatic control of chemical processes this ratio probably will drop. The computers even now can control complex processes and can "learn" (i.e., can alter their programs) based upon the results of the production runs. Consequently all of the conditions are favorable for a rapid and widespread application of automation in chemicals. Most of the effects will probably be felt within the next twenty-five years. The so-called technological unemployment in the chemicals industry that results from automation will be offset many times by the requirements for skilled and scientific personnel in the other industries which service chemicals.

As a final substantive point I should like to mention something that ought to be of particular interest to persons interested in resources. Mr. Stevenson's paper lists the raw materials from which chemical technology draws its products. Note those raw materials well: coal, limestone, corn cobs, oat hulls, and other agricultural by-products. The list could be expanded, but the important point is that these raw materials are low value, plentiful, and in some cases are waste products. Chemical technology thus forces us to revise our estimate of what constitutes the resource base of our economy. Products of chemistry are better and cheaper substitutes for products which come from more scarce natural resources such as our metallic deposits. Furthermore—to borrow a very useful phrase from Mr. Stevenson—we are now concerned with "molecular engineering." Molecules are the building blocks for the new appraisal of resources; no longer are we concerned with the simple mechanical transformation of, say, a metallic ore to a finished metal. Either by direct substitution of a chemical product for the finished metal, or by applying direct chemical reduction to hitherto submarginal deposits, we have suddenly expanded the resource base many times.

From these developments, which are truly revolutionary in their economic consequences, I would like to draw two conclusions briefly. First, the image of an exploding population which presses on a fixed resource base is essentially a myth in the long run. Economic growth will not be limited by a shortage of natural resources. To hoard resources unthinkingly in the name of conservation is to misunderstand completely the nature and scope of technological change. In the short run, to be sure, a shortage of coking coal or high-grade iron ore may impede growth in a given country, but over longer periods such problems will be solved efficiently.

Second, as we face the challenge of accelerating economic growth in the underdeveloped lands, let us recognize an important implication of the phenomenon of "molecular engineering." It will accelerate growth substantially if the underdeveloped countries can import our chemical techniques for deriving useful products from their low-value resources, rather than trying to duplicate exactly our own existing technologies; these have been shaped by the history of our economic development and by our particular resource endowment. Engineers and economists would be well advised to study the possibilities of adapting chemical technology to the production needs and resource characteristics of the underdeveloped countries. That might prove to be our most valuable export to them.

RICHARD L. MEIER

The World-Wide Prospect

LET ME RESTATE for my own purposes the most significant role
that chemical technology plays in the conservation of natural
resources. Among all forms of technology it has done the most
toward finding substitutes for the resources that are rapidly
being depleted. These substitutes are synthesized from the
more abundant resources whose exhaustion seems much more
remote. Second, chemical technology is capable of reforming
the molecular structures that go into ultimate products in
such a fashion as to fit ever more closely the uses to which these

RICHARD L. MEIER is a chemist whose studies for the past twelve
years range over various problems lying between the natural sciences and
the social sciences. He is a lecturer in the School of Natural Resources
and research associate in the Mental Health Research Institute, a be-
havioral science group in the Medical School at the University of Michigan. He received his doctorate in organic chemistry at UCLA in 1944,
did industrial research in high polymers and petroleum chemicals, was
executive secretary of the Federation of American Scientists in Washington, D.C., then Fulbright Scholar at Manchester University, and until
recently assistant professor of the social sciences in the Program of Education and Research in Planning of the University of Chicago. He is the
author of *Science and Economic Development* (1956) and *Modern
Science and the Human Fertility Problem* (1959). His present research
is concerned with industrial planning for developing areas, and with the
growth aspects of organization theory. He was born in Kendallville,
Indiana, but grew up in the heart of the Corn Belt.

139

products are put. As a result, substantial reductions can be made in the strain upon the resources that are needed to maintain a given level of living. Illustrations of the way in which these functions of chemical technology are fulfilled have already been provided by Dr. Stevenson. They were limited, however, to discussions of the American scene, where supplies of critical natural resources are still more abundant than for any other continent. Technology, like science, has universal application, so we may ask what shall be the role of industrial chemistry in developing parts of the world where conservation is even more crucial?

Some of the very large increases in manufacturing capacity for basic chemicals predicted by Dr. Stevenson as necessary to meet growing American needs will probably be located outside the boundaries of the United States. We are already accustomed to accept this situation for aluminum and other refined nonferrous metals, coal tar chemicals, paper pulp, partially refined sugar, etc. It seems highly likely that in the next several decades the production of acetylene, polyethylene, polystyrene, synthetic fibers, low carbon steel, synthetic alcohols of all kinds, detergents, antibiotics, and glass, to name a few major items, will be built up overseas at points where readily accessible raw materials will be drawn upon. We may import some of the output by the carload and tankerful in order to meet peaks in our own consumption, but most of it is expected to be in the form of manufactured goods of a labor-intensive character. The bulk of the manufactured products are likely to be quite simple, for example, knitted goods such as tee shirts and waist-high tights and footwear assembled from plastics, rubber, fabric, and even leather, but some will be as complex as automobiles and portable television sets.

Current reports indicate that the chemical industry is again ready for overseas travel, and its progress may be not unlike that experienced by the textile industry over the three centuries that it has been regionally specialized. Textiles, it may be remembered, were once a group of cottage industries which became strongly localized in Flanders and around Glasgow.

Upon further mechanization the focus moved to Lancashire, and thence to New England. By now it has moved south but even the supremacy of the South is severely threatened by the lower costs and rapidly improving quality in Japan, Hong Kong, and India. The prime centers of the chemical industry have moved in the course of a century from England to the Rhine Valley to the Delaware Basin region to the Gulf Coast. Nowadays there are weekly announcements of participations of American firms in overseas projects, and also of the acceptance of contracts by American chemical engineering and construction firms for setting up plants in foreign countries. The growth in Canada is perhaps greatest of all, but it is so closely interwoven with the chemical industry in the United States we must consider it to be an integral part of the continental chemical production system.

To the south we see a broad-based, widely distributed chemical industry developing in Mexico, with heavy exports of pharmaceuticals and sulfur products already begun. Puerto Rico is building an integrated petrochemicals complex, and a miscellaneous assortment of small fabricating enterprises, the bulk of whose output will go onto the North American market. Venezuela, however, has the resource structure and the location that should enable it to achieve world supremacy in bulk petrochemicals over the next two decades, even though it has had a slow start up to the present time. Trinidad, of course, is trying to carve out a special niche for itself in this and closely related areas of chemical technology within the British Commonwealth of Nations. Argentina and Brazil are making arrangements for internal manufacture in order to save foreign exchange.

In Europe the Common Market, the stabilization of currencies, and the freeing of exchange controls have together acted as a stimulus for large increases in capital-intensive industries, especially those employing chemical processes. Every new finding of natural gas in France, Italy, and Austria is already dedicated in large part to these industries, as is the cracking gas from the newly expanded oil refineries. If the current full-scale tests of the equipment for transporting liquid methane to European harbors is successful, we may expect to see gas replace coal at many points in these economies; in this event the growth rate of chemical industry in particular would be boosted. The Europeans will of course be heavily involved in developing their own markets, but more and more they have to export to live so that they are already concerned with catering to the needs of the rest of the world. Americans may expect to receive from them a long list of extremely high-quality products, as well as surplus by-products which do not seem to fit into their economies, as has happened in the past with naphthalene and cresylic acids. The Europeans will also be exporting chemical technology to the rest of the world as a technical service, and in this activity they will be in direct competition with American firms.

In the Near East quite a few modern installations are coming into operation, but they are limited to products which seem to be basic to the early stages of economic development—cement, fertilizer, oil refining, metal working, water processing, etc. Only in Israel, and to a lesser extent in Turkey, are there signs of any initiative which would lead to the establishment of specialized chemical manufactures. The planners for Iran have incorporated some small-scale facilities in its current development plan.

In the Middle East, India is the dominant economy. Its planners are also emphasizing the heavy chemicals but in another decade they may be ready for some very large-scale, consumer-oriented developments which are built upon the industrial foundation supplied by the steel mills, power plants, textile factories, and research institutes that are now being completed. If any dependable sources of natural gas should become available, either through discovery or overseas transport, the potentials in plastics and synthetic fibers for the masses should be realized much sooner.

In the Far East the picture is dominated by the rapid transformation of Japan into a society capable of employing the most advanced technologies that have so far been created. The soil and mineral resources in Japan have already been exploited to virtually their fullest extent, so that future expansion must be based upon assisting the outlying territories, such as Alaska, Indonesia, the Philippine Republic, and Malaya, in making more intensive use of their natural resources with Japanese organization retaining a share of the output as payment for their effort.

Despite the fact (or perhaps because of it) that Japan has only its human resources to rely upon, it is moving exceedingly rapidly into plastics, fibers, biochemicals, pharmaceuticals, and the extremely refined metallurgy that is now so necessary to modern electronics. Japanese technology is still dependent upon transfers of experience from the United States and Europe, but a backlog of completely new developments of their own is being built up. I am particularly impressed with their ability to work out the troubleshooting problems in the mass culture of algae, an area where Dr. Stevenson's own firm was responsible for early pilot plant work. It appears that a completely new technology, which exploits photosynthesis much more efficiently and must depend upon unusual biotechnical and chemical engineering methods, will diffuse into the under-nourished urbanized areas of the world from Japan. The traditional Japanese diet has been analyzed scientifically with extraordinary patience and skill so that a food technology, used here in the true sense of the word, has evolved from an assortment of crafts. These food-handling processes, particularly the protein syntheses, are much more suitable for feeding the multitudes of Asia and Africa than those developed in Europe and America. It is only by employing such techniques, or their successors, that an adequate diet can be provided to the new city dwellers in low-income territories.

Both in Europe and Asia the Soviet Union is rapidly building up the synthetic rubber, plastics, and synthetic fibers industries. The production of such materials can now be expanded quickly because a good foundation of heavy chemicals and coke-oven by-products is already in place. Although the internal demand for consumer products using chemical-based raw materials is huge, it is expected that imbalances in industrial additions, and occasional inventory accumulations, will be thrown onto world markets. These threats, once exercised, will tend to hold down excess capacity in chemicals and intermediates for the world at large.

Economists whose ideas of industrial policy for economic development are drawn from economic history are prone to misjudge the place of relatively recent technology in the plans for industrial promotion, and the newer chemical technologies are perhaps most often prematurely dismissed from consideration. Certain industries have been associated with low levels of education and skill and have therefore been recommended for the first stages of industrialization. In this category are textiles, garments, shoes, leatherworking, foodpacking, woodworking, ceramics, iron casting, wrought iron fabrication, and wire products. Although circumstances have changed to some extent, it is still usually felt that the newly industrializing countries cannot stray too far from the paths that their predecessors have trod. Modern chemical industries tend to be left out of recommendations to such countries, perhaps because the precedents have not existed.

I have recently had the opportunity to analyze some of these aspects of industrialization in Puerto Rico, a commonwealth that has made remarkable progress over the past decade and has astonishingly complete records. Let me point out first that industrial chemists and economists would consider *a priori* that Puerto Rico is a most unlikely site for chemical technology because it has none of the natural resources, such as cheap hydroelectric power, oil and gas wells, or salt mines, which have attracted chemical facilities in the recent past. Nevertheless, it is worth while reviewing the experience in Puerto Rico.

In the first nine years of Operation Bootstrap, for which complete data on 432 plants are available, it was noted that the most traditional industries for the first stages of industrialization—food processing, yarn making, weaving, knitting, dyeing, garment making, leatherworking, shoe making, furniture making, and the like—were more often unprofitable as private enterprises than the average. This group made only 10 to 15 per cent a year on the capital invested before taxes. During the same period the real moneymaking industries were women's undergarments, paper products, electrical and electronic equipment, cement, pharmaceuticals, and plastics molding, more or less in the order named. Their profits were averaging 25 to 30 per cent a year when calculated on the same basis. The most successful installations were usually one stage beyond a strictly chemical operation such as electroplating, pulping, polymerizing, dyeing, or synthesis. Thus women's undergarments—most factories manufacture bras—require rayon, nylon, and rubber thread and pads.

Information regarding production-per-man-day, a still more significant index than corporate profits, suggests that the newer technologies were responsible for very high labor productivity. Several case studies indicate that the quality of the management normally associated with the new technologies may be more responsible for high profits and high productivity than the capital invested per worker, but these observations merely reflect the standards which are set by the technology. Chemical industry managers, for example, have to be unusually sensitive to human adaptation to equipment because a single error could cause a costly shutdown of the whole plant while elsewhere it ordinarily results in a short series of defective components or assembled products.

The Puerto Rican experience does not seem to be seriously contradicted elsewhere in industrializing territories where the statistics are less comprehensive. We may anticipate that wherever labor-intensive features are involved in preparing chemical products for the American consumer the industry will be susceptible to entry by foreign manufacturers. When a large enough output has been reached in the more distant regions, the chemical plants needed to supply them should begin to sprout in overseas areas, and not necessarily in the same regions as the succeeding stages of manufacture that use chemical inputs.

A natural complementarity in chemical technology, and perhaps for other research-based technologies, then suggests itself. The technology itself will, more likely than not, be created in the United States, starting from fundamental science and advanced engineering experience. When the production system has shaken down, and all the stages to the consumer have been elaborated and demonstrated to have tolerable costs, the new plants will be built. The first, second, and third versions quite probably would be built in the United States. Perhaps the fourth version would be built in Europe, a fifth in the United States, and a sixth in a place like Japan. In that case the Americans would export:

the initial products and components which open up the overseas demand, and provide models for local fabrication and finishing;

the technological "know-how" through blueprints, reports, patent rights, and advice from professional consultants;

the most critical features of the equipment and facilities;

trained technologists, managers, and sales organizers;

troubleshooting services, quality control consulting, and improvements in instrumentation for established operations;

most of the product and process improvements.

And in turn we would buy:

a wide variety of fabricated materials;

some low-cost bulk chemical manufactures, largely by-products;

a series of pleasingly novel designs and styles that result from a new technology interacting with old cultures.

The American chemical industry seems to be of two minds about these trends. It is an industry which stands forthrightly for tariff protection, even though exports of chemicals and products of chemical technology, however classified, are many times greater than imports. This position cannot be explained in terms of international economics or self-interest but perhaps mainly by the early experiences of the industry when it was fighting for survival against the international cartels. At the same time most American firms have willingly entered into "participations" with foreign firms which involved a transfer of their technological experience to overseas sites. The industry has also co-operated with the government in programs for training foreign technicians, in the setting of international standards, and has on its own initiative published widely the results of its research and development efforts.

The outflow of technology, equipment, and personnel and the inflow of finished products, styles, and concepts lead to an amelioration of conditions overseas and to an enrichment of our own society. They contribute to our primary international policy, that of establishing and maintaining a global system of open societies. The scarce resources in the world should be used more efficiently as a consequence of this exchange. Therefore I can see nothing ominous in the trends that have been identified here.

v

NUCLEAR ENERGY

Willard F. Libby:
TOWARD PEACEFUL USES OF THE ATOM

Philip Mullenbach:
GOVERNMENT PRICING AND CIVILIAN REACTOR TECHNOLOGY

E. Blythe Stason:
HUMAN RESOURCES IN AN ATOMIC AGE

WILLARD F. LIBBY

Toward Peaceful Uses of the Atom

NUCLEAR ENERGY first entered man's consciousness with the equation derived by Einstein from consideration of the laws of simple relativity. The equation said that matter in fast motion exhibited an extra inertial mass and that consequently mass and energy might be interconvertible. It equated one gram of mass to 9×10^{20} ergs or about the energy released by the explosion of 20,000 tons of TNT or the combustion of some 3,600 tons of coal. Because one gram of matter is so small an

---

WILLARD F. LIBBY was a member of the United States Atomic Energy Commission from 1954 until June 1959, when he resigned to re-enter academic life as professor of chemistry at the University of California at Los Angeles. He was still a member of AEC when this paper was prepared and presented. During the war years he left teaching to serve as chemist with the Manhattan District project at Columbia University. From 1945 to 1954 he was professor of chemistry at the Institute for Nuclear Studies and Department of Chemistry, University of Chicago. He was a Guggenheim Memorial Foundation Fellow, 1941 and 1951, and continues to serve as an adviser on fellowships to the Guggenheim Foundation. He has made fundamental contributions to the peacetime uses of atomic energy and is especially well known for his work with the isotopes of carbon and hydrogen. Archeological specimens containing carbon are dated on the "carbon fourteen calendar" first prepared by Mr. Libby. His work with tritium, the heavy isotope of hydrogen, has opened new vistas for research in meteorology and ground water. The

151

amount it was clear that an enormous potential source of energy might be locked in the matter of which we consist. This idea first appeared in the early 1920's in the form of Einstein's basic equation: $E = Mc^2$ in which the energy E, is in ergs; the mass M, is in grams; and c is the velocity of light, namely 3×10^{10} centimeters per second.

Professor Gilbert Newton Lewis, of the University of California, was one of the very first people to put this idea to practical application by suggesting that the stars derive their energy principally from reactions of the atomic nuclei which convert mass into energy. At the time it was not known just which types of nuclear reactions were most likely to yield the energy involved but we now know that the fundamental idea is correct and that this indeed is the principal source of stellar energy.

Nuclear energy thus is a very old phenomenon; in a real sense it is the source of nearly all energy because most of the energy we use comes by one means or another from sunlight. The sunlight, due to the photosynthetic process, is able to store energy in chemical forms and give us coal and oil and the fossil fuels, but since the sun is a star and derives its energy from nuclear reactions, nuclear energy is the source of the sunlight and thus of fossil fuels. Now the problems we face in putting nuclear energy directly to work here on earth are many, and I would like to center my discussion in this paper around the problem of the nuclear resources as we now know them and the possibilities of their future development.

Some sources of energy, such as the tidal energies, are not nuclear in origin unless we relate mass itself to the atomic

---

more recent among his numerous professional honors include the 1956 American Chemical Society award for nuclear applications in chemistry; the Willard Gibbs Medal Award by the American Chemical Society, 1958; and the 1959 Albert Einstein Medal and award for outstanding contributions to scientific knowledge. Mr. Libby was born in Grand Valley, Colorado, in 1908, and received his B.S. and Ph.D. degrees from the University of California.

nuclei, which would not be too large an exaggeration since most of the mass of the atom—well over 99 per cent of it—is in the atomic nuclei. The energy of the tides and of all gravitation is not derived from nuclear reactions directly and is in a partial sense non-nuclear in origin.

At the present time, the only nuclear raw materials which promise to be useful in civilian power are uranium, thorium, deuterium and lithium; some two or three other less important materials may come along in due course. Today natural uranium is the prime source for civilian nuclear energy. This is due to 0.71 per cent of the isotope uranium-235 which is contained in ordinary uranium and which is readily fissioned by thermal energy neutrons. Uranium-238 and thorium-232 require high velocity neutrons in order to fission but they are secondary sources also because they absorb neutrons to give new fissionable material, namely: plutonium-239 in the case of uranium-238, and uranium-233 in the case of thorium. The point is that a significant fraction of the neutrons produced in the fission reaction itself are of too low velocity to efficiently fission uranium-238 and thorium-232 so these substances are fertile but essentially nonfissionable in the ordinary chain-reacting system.

The first chain reacting pile which was built by Enrico Fermi in 1942 consisted of pieces of pure uranium interspersed with blocks of pure graphite in a cubic lattice with the lattice spacing somewhat shorter than the distance required for the neutrons formed in the fission of uranium-235 to slow down from their original high velocity to something like thermal velocity at which the probability of absorption in uranium to cause further fission is high. Air was blown through to carry off the heat generated in the uranium fuel by the recoiling fission fragments and in the graphite by carbon atoms recoiling from collisions with fast neutrons. A considerable fraction of the neutrons were actually absorbed by the uranium-238 to form plutonium-239, the yield being something like one atom of plutonium-239 for every uranium-235 fissioning. In other reactors using uranium enriched in uranium-235, thorium can be incorporated and converted to uranium-233. Thus uranium-238 and thorium-232 can, by neutron absorption, be converted into plutonium-239 and uranium-233, respectively, which are about as readily fissionable with thermal neutrons as uranium-235 and thus are nuclear fuels. This is the basic set of principles on which nuclear energy from fission rests at the present time. Now the resources are, of course, uranium and thorium and the moderator materials such as graphite necessary to decelerate the fission neutrons. An important alternative moderator material is heavy water, that is the water with the mass 2 hydrogen isotope, deuterium, in place of the ordinary hydrogen in the water molecule. The Savannah River reactors have heavy water as moderator. The Hanford reactors use graphite as moderator. Both are designed for the purpose of producing fissionable material—plutonium-239, though both can also produce tritium by the irradiation of lithium. Tritium is useful in the fusion type of reaction which is the other type of atomic energy we have to consider.

In the early studies of nuclear energy, it was shown by weighing the atoms that both the light and heavy atoms were heavier per given number of particles than the intermediate atoms around iron. It was therefore conceivable, in theory, that according to Einstein's basic equation fusion of the hydrogen isotopes to form helium or even heavier atoms could release energy. This is called the fusion-type of nuclear energy. The other possibility was that the heavy atoms such as uranium could, by splitting, form materials of lower specific mass content in the middle of the periodic table. This is the fission type of energy. Now, the principal raw material for fusion is deuterium, the same raw material used for moderator in certain types of fission reactors. One also can use lithium to make tritium, and tritium itself is a readily fusable hydrogen isotope, so uranium, thorium, deuterium and lithium are the fundamental raw materials on which nuclear energy is based.

Now let us consider the abundance of these materials. Before the atomic age, which began in December 1938, when

Professor Otto Hahn in Germany discovered the fission reaction, uranium was a relatively unused chemical. Its principal use was as a coloring agent in the ceramic industry—the yellow and orange glazes obtained with it are well known. Also, since radium is the radioactive decay product of uranium-238, which has a lifetime of some 6 billion years on the average, uranium ores are the only source of ordinary radium used in hospitals and elsewhere. But this business was a small one and when it became necessary to produce thousands of tons of uranium to build chain reacting piles, larger quantities had to be found.

At the time the only available source of any magnitude in the Western World was the Shinkolobwe deposit in the Belgian Congo and it was through the hearty co-operation of the Belgian government and the Unione Minière de Haut Katanga, and very particularly through the efforts of Sir Edward Sengier, that this material was made available to the Allies. For years Belgium was a principal source of the uranium available to the United Kingdom and the United States. We in this country had mined a little uranium ore in Colorado, mainly as an incidental to development of vanadium, but it had not developed to the magnitude of the Congo production.

As a result of the Atomic Energy Commission's activities and those of friendly nations, AEC's situation is quite different today. The joint efforts of the United Kingdom, Canada, and this country, together with those of Belgium, South Africa, and Australia, have developed additional sources so that now uranium is available in what is virtually an excess when viewed in the short range. We purchase about 30,000 tons of uranium oxide per year and we now see no problem in meeting our requirements during the next ten years. Since most of the present reserves have been developed only in the last few years, additional reserves will be developed as the market for uranium increases.

Experience has indicated very strongly that large additional domestic deposits will be found if they are needed. Canada also has discovered great resources and it is clear that large areas of the world have not been very thoroughly explored.

In particular, Africa probably has considerable resources as yet undiscovered. The French have found fair-sized deposits in metropolitan France, and there are reasons to believe that Russia has extensive deposits within her borders and in the satellite countries. Eventually, it may be necessary to turn to the low-grade shale and phosphate deposits, and here we have a tremendous source of uranium—a source that could supply the world's nuclear fuel requirements for many decades.

Although the prospective demands of the next few years can be satisfied by presently known deposits and large new discoveries are quite conceivable, the longer run outlook is less clear. The uncertainty comes chiefly from the possibility that a considerable part of the expected growth in United States and world power needs will be met by installation of nuclear power plants. If this happens we can envisage perhaps 100 million kilowatts of electric power based on nuclear energy in this country. The requirements for uranium for such a program become quite large. To suggest *how* large, let me give you just a few numbers. Roughly speaking, a kilowatt of electricity generated continuously over a period of one year requires the fission of about one gram of uranium-235. Now each ton of uranium contains about six and a half thousand grams of uranium-235. Therefore, 100 million kilowatts, if it were kept busy twenty-four hours a day for a whole year, would fully utilize the uranium-235 contained in 15,000 tons of uranium. However, it isn't possible to run reactors to the point at which all the uranium-235 is used up. For technical reasons, it is necessary to remove the uranium fuel elements for reprocessing at a considerably earlier period and this has an important effect on the uranium inventory. To keep a given reactor running there will be good uranium in the used fuel elements, a set of spares will be on hand or in fabrication, and there will be a substantial amount of uranium in the diffusion plant where uranium-238 and uranium-235 are being separated to make enriched uranium fuel.

What this adds up to is that inventories and pipelines contain several times as much uranium as is consumed annually.

During the period in which many new reactors are being built, the principal requirements for uranium will be to supply the initial charge and fill the pipelines. Therefore, we can see that in the instance of a large increase in the amount of atomic power the present 30,000 tons per year would not be adequate. The question really is whether the amounts of uranium that would be required for a world-wide atomic power industry of considerable magnitude—say 50 per cent of the total installed electric power—could be supplied throughout the next century by the mining of natural uranium. The differences of opinion on this question are very great and it is by no means clear that it can be supplied by future discovery of deposits of uranium. The question is somewhat analogous to the perennial debate over the exhaustibility of oil resources.

It is possible, therefore, that we may be short of uranium if we depend upon the burning of natural uranium or of uranium-235 derived from it as our only source of nuclear energy.

There are a number of other possibilities. First, let us consider the supply of thorium, heavy water, and lithium. All three of these materials are present in quantity and undoubtedly can be made available in adequate amounts. Extensive deposits of thorium in India, Brazil, and Canada are known to exist, and in this case there is only one isotope—it is all thorium-232. True, it is like uranium-238 in requiring neutron irradiation to make fissionable fuel—uranium-233—or the use of very energetic neutrons to directly fission it.

Ordinary water contains about one part in 6,700 of heavy water so there is no conceivable shortage here. Lithium is very abundant since small amounts are to be found in many igneous rocks. The lithium-6 isotope, which can be converted into tritium for use in fusion reactions, constitutes 11 per cent of the ordinary lithium and, therefore, the supply is quite adequate. Now, of course, there is also nuclear energy from fusion. We have a program called the Sherwood Project which is devoted to the taming of the hydrogen bomb. The problem is to learn how to make the fusion reaction proceed in a slow and orderly fashion rather than in an explosive way and thus to supply nuclear energy from fusion. If we should succeed in taming the fusion reaction—and it seems that it will be many years before this attempt is successful—we would, of course, have a virtually inexhaustible supply of raw material in the waters of the earth.

So, in the short range uranium is available in adequate supply, but in the long range the questions of its adequacy are not clearly answered, and it might be that uranium sources would be depleted and other steps would have to be taken.

Let us now consider what these other steps might be. We are trying to learn how to use thorium and the uranium-233 derivable from it. Also, we are considering the possibility of better utilizing our uranium supplies. Plutonium-239 is produced by neutron absorption in uranium-238 just as uranium-233 is produced by neutron absorption in thorium-232. Both of these isotopes are thermally fissionable like uranium-235 and are good atomic fuel though we have problems in learning how to burn them, particularly in the case of plutonium-239. A vigorous effort is in progress to surmount these difficulties. It is theoretically possible to convert uranium-238 or thorium into fissionable form with an efficiency which either equals or exceeds that necessary to replenish the uranium-235 burned up. We call such devices breeder reactors and every country in the nuclear energy business is trying to make them. This for fear that the uranium supply will prove to be inadequate. We now have under construction at the National Reactor Testing Station, at Arco, Idaho, a $30 million breeder reactor installation to test the possibilities of this reaction, and the Power Reactor Development Company in Detroit is building a similarly large installation for the same purpose. In addition, a considerable effort has been under way for a number of years at the Oak Ridge National Laboratories for another type of breeder.

All of these experiments will help answer the question of whether it is practicable to convert uranium-238 and thorium-232 into thermally fissionable materials. If so, we could feel sure that uranium and thorium supplies are adequate and that nuclear energy from fission will not suffer in the foreseeable future from lack of raw material. The amount of uranium and thorium, if we could burn it all, is entirely adequate to meet our energy needs for a long time. For example, the amounts of these materials in ordinary granite makes granite several times as energy-rich as coal, and if we can find a way to get these materials out at a reasonable cost, the supply is vastly increased.

If breeders will work it will be possible to burn all of the uranium instead of 0.71 per cent of it, and possible also to utilize thorium. At this time it is not possible to say how certain success is in this area. We believe that success is likely enough and the need probable enough so that we should invest both public and private funds in breeder reactor experiments and development efforts.

The great question I mentioned before: Can we discover more uranium? No one knows the answer to this but it seems likely that we can, and the possibility of a Mesabi range of uranium ore is one which we have often speculated about, especially its effects on the economic aspects of nuclear power as we now know it.

The third question is, of course, the question which Sherwood will answer: Will it be possible to tame the hydrogen bomb and get useful energy from it?

An entirely different and much newer development in the nuclear energy field may have great possibilities. This is the project known as Plowshare whose purpose is to put nuclear explosives to non-military applications. This possibility has been discussed for a number of years, even as early as World War II. The advent of hydrogen bombs greatly increased this possibility because of the comparatively low cost for a given energy release and the small percentage of fission products formed.

Most of the ideas for peaceful uses of nuclear explosives fall into three broad categories. These are related to three effects: (1) a nuclear explosion can break or shatter and move large amounts of material; (2) it can heat materials and this heat can be trapped within the earth for later usage; (3) it liberates a very large number of neutrons in a very short time and these neutrons can be put to use.

On the first effect, the obvious applications are to move earth so as to make harbors, canals, and other useful civil engineering projects or to remove overburden for strip or open cut mining. It should be possible to cheaply remove considerable quantities of earth lying over ore deposits. Rock could be shattered so as to make mining operation more economical or to create catch-basin aquifers for the storage of water underground. In January 1959 there was a conference in Dallas, Texas, on the utilization of atomic explosions in oil shale. The conference was held through the co-operation of the Bureau of Mines and the Atomic Energy Commission and its contractor, the University of California Lawrence Radiation Laboratory at Livermore. Some three or four dozen industrial companies were represented and many of the leading geologists and oil scientists from all over the country took part in a two-day discussion of the possibilities. The question was the feasibility of using nuclear explosions to help in the extraction of oil from oil shale by breaking up and heating great masses of rock. A principal question was what would happen if we were to detonate, say, 300 kilotons of nuclear explosion in the Green River oil shale deposit which lies in Wyoming and Utah and Colorado, and has total reserves of at least several hundred billion barrels of oil. On the basis of previous experience at our proving ground in Nevada we would expect 300 kilotons fired in the Nevada rock to result in about 88 million tons of crushed rock, and 35 million tons of broken rock from caving. An equivalent amount of broken shale might have 20 to 40 million barrels of oil in it. The Piceance Creek Basin of northwestern Colorado was suggested as being particularly appropriate for a test of about 10 kilotons. It was suggested that a device lying below the rich oil shale beds of the Mahogany ledge could be fired beneath an overburden of about 970 feet or more. Gas and oil would be produced from the organic matter by the thermal energy of the explosion. The actual quantities of the products will depend upon the temperature history of the shales surrounding the explosion, the grade of the shale, and other factors.

If one-half the energy released by the device were utilized in heating shale to a retorting temperature of 900°F, this would be enough to produce about 10 million cubic feet of gas and 15,000 barrels of oil. However, because of the extremely high temperatures and pressures produced in the explosion it is unlikely that the quantities of gas and oil actually formed bear any direct relationship to these calculated quantities. Since the Green River oil shale contains silicon it is probable that the radioactive fission products will be trapped in an insoluble fused rock as occurred in Nevada with the Rainier test.

The second category of uses in the Plowshare Project is the trapping of heat. The accompanying figure shows the distribution of heat from an underground shot which was fired in the volcanic tuff formation in the Mesa in Nevada on September 19, 1957. This profile of the temperature distribution was determined about five months after the detonation. About half of the energy was left trapped in the rock. The rock was porous and wet and as a result none of the temperatures are above the boiling point of water, but presumably the temperatures originally were much higher and had the rock been dry undoubtedly would have stayed there and might have been used to make useful steam by pumping in water. The steam so generated could possibly have been turned into useful power. In the Rainier shot to which these data refer, a 1.7 kiloton explosion in volcanic tuff made some 700 tons of fused rock which had been heated from 1200° to 1500°C. The energy required for this amounted, in itself, to about one-third of the total energy released by the device. In addition to the energy which fused the 700 tons of rock, energy released by the shock wave heated the rock beyond the 55-foot radius vapor bubble which the bomb created. This bubble had a skin 4 inches thick consisting of the 700 tons of fused rock. When it cooled, a glass was formed which sealed in the radioactivity. We found from studies of the site that the bubble must have developed a steam

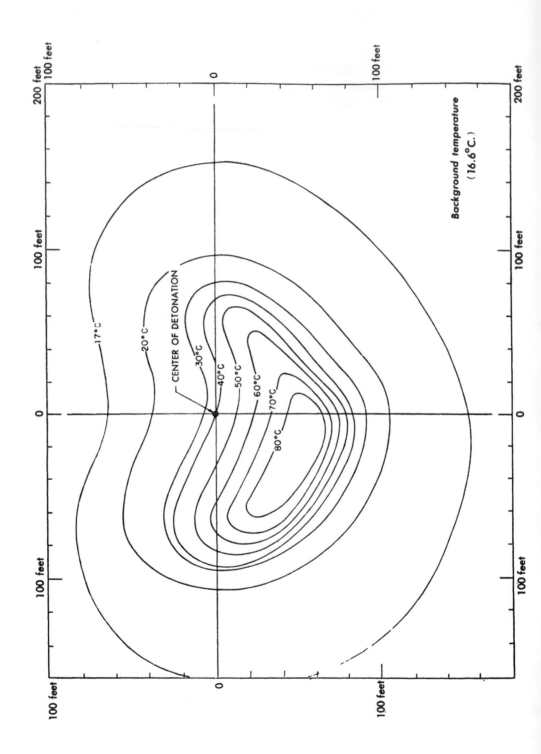

CENTER OF DETONATION

17°C
20°C
30°C
40°C
50°C
60°C
70°C
80°C

200 feet
100 feet
0
100 feet
200 feet

100 feet
0
100 feet

100 feet
0
100 feet

Background temperature
(16.6°C.)

pressure internally of about 40 atmospheres, about equal to the weight of the 800 feet of overburden, and that after cooling a minute or two the crushed rock above it cracked the bubble like an egg shell and fell into the cavity, intermingling with the shell fragments so that the radioactivity is now found in fragments of dark colored flint-like glass easily distinguishable from the other material because of its color and general appearance.

Now let us consider whether we can make atomic explosions sufficiently cheaply and contain the heat so that it might be possible to make economic nuclear power from buried nuclear explosions. The economics of nuclear explosions are such that it is unlikely that kiloton-sized explosions will be able to produce energy cheaply enough for nuclear power to be economic. However, buried hydrogen bomb explosions producing energies equivalent to millions of tons of TNT may be another story. Per unit of energy released, they are much cheaper than atomic bombs in the kiloton range, which, of course, are in themselves much cheaper than ordinary chemical explosions. In fact, the actual cost of these large devices is such that as far as the energy is concerned there is little doubt that we should be able to make economic nuclear power if we could trap and make steam out of the energy. And so it may be that this method of taming the hydrogen bomb may succeed before the Sherwood Project approach of making the reaction go slowly. In any case, this is one of the objectives of the Plowshare Project—to try to tap the energy of the hydrogen bomb to make nuclear power.

Nuclear explosions liberate large quantities of neutrons instantaneously, and these may be put to useful purposes in making isotopes which normally require extensive irradiation in chain reacting piles. This is the third general category of Plowshare applications. A nuclear explosion lasting a tiny fraction of a second involves an amount of fission or fusion which is comparable to the total involved in a good-sized chain

Opposite: Temperature in vertical plane surrounding the Rainier detonation point, measured five months after detonation.

reacting pile operating for several months or a year or so. So, essentially, there is the possibility of generating almost instantaneously the isotopes which would be made in a year or so in a large reactor. Of course, the problem is to recover them from the debris of the explosion in a sufficiently inexpensive way. There seem to be definite possibilities of doing this.

It might be helpful to describe in a little more detail the four experiments which have been conducted on the underground detonation of nuclear devices. First was the Rainier event of September 1957. It took place in a room 6 by 6 by 7 feet at the end of a tunnel driven into the side of a mountain— really a mesa—in the Nevada test site. The geology of the mountain was characterized by a cap of welded tuff about 250 feet thick overlying several layers of bedded tuff to a depth of about 2,000 feet, all of which lies on the basement rock of dolomite. The tuffs vary from unconsolidated to consolidated. The tunnel itself was in a consolidated bed so that no shoring up was required. The tunnel section was 7 by 8 feet. The density at the site was 1.9 to 2.0 grams per cubic centimeter with about 15 to 35 per cent water present. Because of this water it was preordained that we would not be able to obtain rock hotter than the boiling point of water no matter what happened. We didn't realize that before the experiment was undertaken. It is strange how simple things are sometimes missed.

The tunnel ended in a room which was 899 feet below the top of the mountain and 790 feet from the nearest point on the sloping face. Actually the tunnel had a spiral ending so that the shock would not be able to fire right back down the tunnel, though it was found later in our fourth underground shot that with proper sandbag barriers a straight tunnel could be used. At the firing point $2\frac{1}{2}$ miles away there was a muffled sound and a little ground shock; scattered rocks were shaken loose from the cap of the hill for distances of about one-half mile on each side of the point of detonation. After the explosion it was found that collapse of the tunnel had occurred only to a radial distance of about 200 feet. Some splintering of the tunnel walls had continued out to about 500 feet from the device. Beyond that distance to the head of the tunnel, some 2,000 feet in total, there was essentially no damage. The fragmented material in the entire length of tunnel beyond 400 feet from the bomb amounted to a few cubic yards. At 1,100 feet from the bomb, however, there was a shift along a bedding plane of about one-third of a foot.

As far as could be determined, the radioactivity was completely contained, although very small leaks to the surface could have evaded detection. No radioactivity was discovered inside the tunnel even though the search used the most sensitive techniques known for the radioactive rare gases, krypton and xenon.

Acceleration and displacement measurements indicated that directly above the point of detonation the shock arrived at 0.146 second after the detonation and that the maximum vertical displacement was 9 inches. The vertical displacement at a horizontal radius of 1,000 feet from a point on the surface directly above the detonation was less than $\frac{1}{2}$ inch. Analysis of all results indicate that a block about 100 to 200 feet thick separated and rose and fell as a unit, with a maximum acceleration on the surface of 5.8 times gravity at 0.186 second after the firing. What happened apparently was that the shock from the bomb for the first 3 feet was sufficiently strong to vaporize the rock and to melt it out for the next 12 feet. As the shock moved outward, the room expanded spherically and reached a radius of 55 feet in several thousandths of a second. At this time the cavity was lined with about 4 inches of melted rock at a temperature of 1200°C to 1500°C, and the cavity was filled with steam at a pressure of about 40 atmospheres which is a little less than the lithostatic pressure of the rock at that depth, 900 feet. The cavity stood between half a minute and two minutes. Much of the fluid rock had time to flow down the sides and drip from the top. You can actually see this taffy-like material lying in the debris. After about two minutes the steam pressure dropped and the cavity collapsed like an egg shell as it had inadequate strength to support the rock which was now fractured out to a radius of about 130 feet. The cavity was filled with broken rock and the caving progressed vertically to a distance of about 386 feet above the point of detonation.

About one-third of the energy of the bomb went into melting the 700 tons of rock which constituted the four-inch-thick egg shell in the bubble. This rock was very hot—about 3,000°F—and as soon as the caving occurred the water in the general surroundings hit it and made steam. The steam distilled out, carrying the heat with it. In this way the whole of the energy trapped was disseminated rapidly over a sphere. Even now when one goes into the Rainier tunnel the rock is still so hot that one cannot hold one's hand on it for very long.

During October 1958, some underground shots were fired. One of the most interesting of these was the Logan shot of 4½ kilotons, as compared to the Rainier 1.7 kiloton yield. The Logan shot was fired also at the same 900-foot depth, approximately, and roughly the same phenomenology occurred, except, of course, that everything was on a somewhat larger scale. This time the tunnel collapsed out to 820 feet. The ground shock was readily felt by observers at 2½ miles.

The last underground shot was Blanca which was fired on October 30 at the very end of the final test series. It again was buried at roughly the same depth as both the Rainier and Logan shots. Whereas both Rainier and Logan did not break through to the surface, Blanca did. The point was that in this case the caving in following the crushing of the egg shell extended all the way to the surface. The shot room was 7 by 8 by 20 feet; the vertical depth was 988 feet, and the nearest point to the surface was 835 feet. On detonation the shock was clearly visible through the effect it had of kicking up dust. A large quantity of the cap rock was broken loose and fell down the side of the mesa. At about 15 seconds a plume of dust was ejected due to the collapse of the initial cavity to the surface. This gave a dust cloud which rose about 1,000 feet and carried a little bit of radioactivity in it. The fraction of radioactivity which escaped is estimated to have been about 0.1 per cent. In the Rainier shot the cavity caused by the collapse of the egg shell also contained a tiny bit of radioactive krypton, one of the gaseous fission products, and in the Blanca shot a roughly corresponding amount was found. There was also a very small but detectable leak of radioactivity into the tunnel. By the way, the tunnel did not collapse much farther out than it did in Logan—that is, out to about 850 feet. However, there was a thrust of about 3 feet along the bedding plane at a distance of 1,500 feet from the bomb and a movement of about one foot vertically on a fault at a distance of 2,700 feet.

The ground shock was felt very strongly by observers at a distance of 5 miles and was noticeable as a rolling effect at 16, but not at 20 miles. At a distance of 2 miles the earth's motion was strong enough to rock automobiles violently from side to side. Thus, we see the general phenomenology of underground shots up to the maximum of 23 kilotons buried at something like 900 feet in the volcanic tuff in Nevada. From the Plowshare point of view there is a wealth of information in these tests.

In order to proceed further with Plowshare it will be necessary to conduct additional tests. When the negotiations in Geneva have reached the appropriate stage we plan to proceed with a shot in the salt deposits near Carlsbad, New Mexico, so that we may test the possibilities of making atomic power and isotopes. We choose salt because it is certain to be very dry; thus the heat will not be degraded in temperature below the melting point of the salt which is 800°C. Finally, the isotopes would be contained in a water-soluble media which would be easy to mine and process so the recovery operations should be cheaper. The unburned fissionable material in an exploded device with yields in the kiloton range is a major fraction of the total fissionable material contained since the critical mass is larger than the energy release and this unburned plutonium or uranium-235 has a considerable value. Therefore, even without isotope production, mining of buried shot debris could be a paying matter. In the 700 tons of fused rock from Rainier there are a few parts per million of fissionable material. Because of the difficulty of dissolving this glass, recovery of the fissionable material is hardly worth while, but if it were in a soluble salt, recovery might be inexpensive. Also, one could economically recover isotopes manufactured by placing various target materials around the device. Among the important materials which can be made in this way, of course, are tritium from lithium, plutonium from uranium-238, and uranium-233 from thorium, and other isotopes made by neutron reactions.

There are two sources of nuclear energy: fusion and fission. Each of these subdivides into controlled release and uncontrolled release and, thus, we have essentially four potential sources of nuclear energy on earth. Of course, all energy has a nuclear origin essentially in that the sun runs off nuclear reactions—the conversion of hydrogen into helium. However, when we speak of nuclear energy we are referring to the possibility of using atomic nuclei to make power here on earth. In this sense there are four potential new energy sources. Of these, the one on which the most progress has been made is nuclear fission from reactors. The ordinary reactor which operated on the fission reaction is close to economic power. In my opinion it is likely that in about five years this will be demonstrated as an actuality in Western Europe and within ten years will be conclusively established in wide areas in the United States.

Of course, we are all familiar with the fact that in the United Kingdom considerable effort is being made to develop gas-cooled fission reactors for the production of nuclear power. The Atomic Energy Commission has become interested in these recent months in the gas-cooled type of reactors also, and we plan to enter into the field soon. However, our principal efforts have been with water-cooled reactors; reactors in which steam is produced either by hot water under pressure which makes steam through a heat exchanger or by water which boils to make steam and carry away the heat, called pressurized water reactors and boiling water reactors, respectively. In work on these two types of reactors we are far ahead of the pack, and we expect that the earliest economic atomic power will come from either the boiling water reactor or the pressurized water-type reactor.

In the longer range, however, water reactors are limited in efficiency because it is impossible to get a really high thermo-dynamic efficiency from reactors cooled by liquid water. The simple point is that water boils at such a low temperature that it is impossible to get it very hot without converting it into a gas. Why not convert it into a gas? That's a good idea, except that water is a very corrosive gas; if you are going to make a gas-cooled reactor you had better choose something like helium or carbon dioxide which is more inert. This is exactly what we plan to do in the gas-cooled type of effort and this is the principal argument for gas-cooled reactors—that they do have potentially higher thermodynamic efficiency. In the long run it probably is not water reactors which will be the most efficient, but possibly gas-cooled reactors or another type of reactor which has been coming along quite rapidly—the organic-cooled reactor. This is the reactor in which the heat is transported by organic fluids similar to some of the less volatile oils, which can be heated to higher temperatures than are obtainable with water at the same pressure. Also, they do not become as radio-active as does liquid sodium in a sodium-cooled reactor we are working on. One difficulty with the organic moderated reactor is that the organic material is decomposed by the radiation of the reactor and this requires replacement which adds to the cost. But considering that pressure vessels are not needed in the reactor and that safety problems are minimized, we have hope for the organic-type reactor. It may be that the present type in which the organic material is both the moderator and the coolant is best, but a type of reactor in which the organic chemical is the coolant and other material is used as a mod-erator may be more promising. Probably the additional knowl-edge we will gain of radiation decomposition will make it pos-sible to control the formation of gums and thus make the reactor more promising.

Also of great significance is the possibility of a chemical reactor and of the use of reactors to produce isotopes. Nuclear energy is important but there are other applications of the nuclear reactions which we should consider as being equally, or perhaps more, important; these are the applications of isotopes and radiation. Isotopes and radiation at the present time constitute a great benefit to mankind. We should never forget the use of isotopes in medicine and industry is at the present time a great benefit even though the total expenditures on the development of isotopes and radiation are hardly one per cent of those which have gone into nuclear power.

We have been trying, in the Atomic Energy Commission, to develop the isotopes and radiation program further but the general enthusiasm for other things has made it rather difficult to do so. At present, however, we have an effort in this direction. Nuclear energy can derive considerable benefits from the incidental production of radiation and isotopes for sale and possibly from the utilization of the reactors themselves to produce chemical reactions. By a chemical reactor I mean a reactor that may be able to make nuclear power and chemicals simultaneously. For example, at Hallam, Nebraska, we plan to build a sodium-cooled reactor in which the coolant sodium becomes radioactive when it is irradiated with neutrons. Therefore, the sodium which carries the heat out of the reactor after going through heat exchanger to form steam still is intensely radioactive with fifteen-hour half-life radiosodium. The plan is to put the radioactive sodium through a series of piping pancakes in an adjoining room in which truckloads of wheat and meat can be irradiated. Simple calculations show that for a few hundred thousand dollars it will be possible to irradiate hundreds of tons of food per day to a level where it can be pasteurized by the radiation at a vary small fraction of a cent per pound of total cost. Hallam is fairly close to Omaha, the large meat producing center, and also is close to large wheat producing areas. We think this may be a good example of a case where nuclear power reactor might be used incidentally in an important way as a source of radiation. For the production of isotopes there are some real possibilities of using nuclear power reactors to make them. We have had very little success so far in interesting the companies building atomic reactors. They seem to not want to bother. However, I think as the uses of isotopes develop it will be necessary either to build reactors to produce isotopes or to interest the builders of nuclear power plants in doing it incidentally to their power production. The principle of by-products is a very old one and it certainly applies to the atomic power business.

The chemical reactor is such an early development it is very difficult to assess, but we now have a proposal for a study of the feasibility of the production of nitric acid by the irradiation of air in an air-cooled type reactor—the air forming nitrogen oxide which is, of course, the precursor to nitrogen dioxide and nitric acid. The possibilities look encouraging. There may be other instances of where chemical reactions may be profitably catalyzed by the intense radiation in atomic reactors.

It is clear that the future is rich indeed because of nuclear energy, and that if we follow the possibilities wisely and assiduously we will be greatly rewarded for our efforts. Wisdom and understanding once again are the determining factors.

PHILIP MULLENBACH

Government Pricing and Civilian
Reactor Technology

AS COMMISSIONER LIBBY already has noted, the outlook for non-military uses of nuclear energy is not wholly a matter of science and technology. He has pointed out that if the new processes are to be widely adopted they must also be economic; that is, able to compete with conventional energy sources in terms of cost and price. In so doing, he has not only rounded out his presentations of future prospects, but supplied me with a

---

PHILIP MULLENBACH is research director of the study being under-taken by The Twentieth Century Fund on the economic aspects of nuclear power policies. He came to this assignment as an outgrowth of his work as research director of the National Planning Association's project on the productive use of nuclear energy, which turned out twelve economic studies, 1955-58, financed by grants from Resources for the Future and The Ford Foundation. During the period 1947-54 he was principal economist on the staff of AEC in charge of preparing classified reports to the President, the Joint Committee on Atomic Energy, and the Commission, on the progress of the atomic energy programs. A long-time civil servant in Washington, first in 1934, he has had experience as an economist in the surplus property program, the War Department in World War II, and the reciprocal trade agreements program earlier. Mr. Mullenbach was born in Chicago, Illinois, in 1912, received his under-graduate degree in economics from the University of Chicago in 1934, and did graduate work at Columbia University, 1936-38.

springboard for this paper, which explores some aspects of the quite extraordinary economics of nuclear power, and particularly the questions of price determination. This comes naturally; price, as you know, is the be-all and end-all of economists. A special reason, however, for concentrating on governmental pricing is to help reveal how pervasively nuclear reactor technology is being influenced by these policies. This technology, it will be seen, does not grow solitary in a laboratory vacuum; it is fitfully subjected to the external pull of public policy and economics.

Let us begin by tracing, without too much detail, what reactor specialists call the nuclear fuel cycle, attempting at various points to relate price policy to the pertinent aspects of the evolving technology. Our questions, though obviously simplified, should be sufficient to reveal the tangled issues confronting the federal agency that must set prices long before either the technology or the economic conditions of supply and demand for nuclear power are known with confidence. As a background to our inquiry, it may be helpful to keep in mind that fuel cycle costs for a typical power reactor using enriched uranium are currently estimated at 4 to 5 mills per kwh, or 40 per cent of total generating costs, 10 to 12 mills per kwh for a plant operating close to capacity.

These are the questions we shall explore: First, how do the various prices fixed by the Atomic Energy Commission impinge on major technical aspects of fuel cycles? Second, how important are the future uncertainties in price setting, as contrasted with the more obvious uncertainties of fuel technology? Third, what seems to be the underlying motivation of government pricing policy and methods? Finally, can we conclude anything useful about the economic wisdom of such pricing practices in the future?

Now, in exploring these matters we start under a handicap: a large part of the most essential cost data required for evaluating fixed prices is classified as national security information and cannot be analyzed publicly. (The same, incidentally, is true of the nuclear explosions described by Dr. Libby.) This means the economics of nuclear power in the strict sense is unknowable. Paradoxically, although power reactor technology is "open," its economics is "closed" to public view.

A typical nuclear fuel cycle begins with the fabrication of natural uranium or enriched uranium into fuel elements that have favorable qualities for sustaining a controlled fission reaction. Fuel elements should permit long irradiation in the nuclear reactor, resist corrosion and damage, as well as have good heat-transfer properties—all these factors being optimized to lowest possible fuel cost. The periods of fuel irradiation and burnup in the reactor may range from a few weeks to many months, but even during long exposure perhaps little more than 1.2 per cent of the total amount of uranium may be burned up, based on current technology with slightly enriched fuel. This is the equivalent of 10,000 megawatt days (heat) per short ton of uranium, or roughly 30,000 tons of coal.

Upon being removed from the reactor, the fuel elements are allowed to stand so that much of the short-lived radioactivity can diminish. Then the fuel elements are chemically processed, to remove the fission products, and to recover both the valuable remaining U-235 and the plutonium that has been formed out of U-238, as described by Dr. Libby. Upon being re-enriched to the level required by the reactor design, the uranium is again fabricated into fuel elements for return to the reactor. Thus the stock of uranium inventory required for a reactor's operation is far greater than the amount actually within the reactor at any given time. This is a fact of both economic and technical importance.

So much for a highly simplified description of the fuel cycle. There are many variations and deviations, of course, that must be ignored for our limited purpose. One of the more important is that by-product plutonium may be used as reactor fuel or turned back to the government at a price.

Under the Atomic Energy Act of 1954 the government continues, as it did under the 1946 Act, to own all fissionable material, though not natural uranium. This introduces an economically strategic factor into price policy, into fuel cycle costs, and into future fuel cycle development. *First,* AEC provides enriched uranium on loan to reactor operators and thus far has chosen to apply a 4 per cent charge on the value of the quantity tied up. *Second,* AEC charges a price for the amount of U-235 actually burned up and the cost of re-enriching the spent uranium to the level required by the particular reactor. The fuel prices, which were fixed on the basis of production costs by AEC in 1955, range, for example, from $9.70 per gram of U-235 for fuel containing 1.5 per cent U-235, a typical power reactor fuel, up to $17.11 per gram of U-235 for uranium enriched to 95 per cent by weight. This range reflects the difference in the cost of producing slightly enriched compared with highly enriched uranium; this difference in turn explains why reactor operators find it cheaper to burn slightly enriched rather than highly enriched uranium fuel. *Third,* AEC pays the reactor operator for having produced any plutonium built up in the fertile material, U-238. The AEC buy-back price of plutonium now ranges from $30.00 to $45.00 per gram, depending upon the amount of associated isotopes that reduce the weapon value of the plutonium. In contrast with its weapon value of $30.00 to $45.00 per gram, plutonium is commonly assigned a value of only $12.00 per gram for fuel purposes in nuclear power reactors.

In addition, since the AEC is the only present source of chemical processing services, mentioned earlier, it makes a charge for performing these functions in government-owned facilities. The service charges established by AEC are for a hypothetical privately owned plant, with an assumed capacity commensurate with future commercial requirements.

The economics of nuclear power in some measure then are being determined by actions already taken, and to be taken, by the AEC as price setter for government materials and services. Recently, I took some nineteen items in a typical fuel cycle and with their corresponding estimated costs, roughly measured what fraction of the total fuel cost was contingent upon prices or charges set by AEC and within its discretion to alter. Briefly, nine of the nineteen items were directly influenced by AEC prices or charges and these items together represented nearly two-thirds of the total estimated fuel costs for the example chosen. In the order of their importance, these fixed prices ranked highest: the 4 per cent use charge, the price of U-235, the price paid for plutonium and the chemical processing charges. Therefore, on top of the large cost uncertainties introduced by the new technology itself, one must add a substantial further uncertainty for government price policy and practices.

Granting that these prices do influence fuel cost, let us ask now what difference AEC prices may make in the direction of reactor development and the allocation—or, possibly, misallocation—of resources. The differences in practice are usually subtle, but a few obvious ones can be cited for illustrative purposes. The 4 per cent use charge, for example, is so low as a rental charge that only a small penalty falls on reactor operators who may carry a large inventory of nuclear fuel. This is the case, for example, of the fast breeder reactor, which necessarily has a very large inventory of fuel in process compared with most other types. Or, again, the comparative advantage of using stainless steel clad fuel elements as against more costly zirconium clad fuel elements is influenced by the 4 per cent use charge. The present low charge favors the less efficient stainless steel fuel element, which requires a greater amount of U-235 to provide the same amount of heat. The allocation of resources does not appear to be the most economic, that is, least wasteful.

But let us take a more difficult example: When the AEC set the charges for chemical processing of irradiated fuel elements it apparently was caught in the dilemma of having to choose between a *low* charge, suggested by the great economies of scale in its own plants, as against a *high* charge suggested by the desirability of getting private enterprise into the business, a laudable objective under the Act. The AEC choice could prove crucial for reactor technology. The effect of a high, "producer's-incentive" charge is to encourage chemical companies to provide this service—though none has been prepared to do so yet. A low charge, on the other hand, while discouraging private suppliers of the service, would have the advantage of encouraging the use of by-product plutonium as reactor fuel, because its recovery would be less costly to the reactor operator. AEC must balance the cost and benefits of such desirable and conflicting objectives in this, as in many another, case. But in this instance plutonium recycle seems a more important goal to serve, again on the grounds of less wasteful use of resources.

One final example of the leverage a particular price practice exerts on technology: If AEC, as it is now doing, sets prices on by-product plutonium which are substantially *higher* than the value of this material as reactor fuel, certain technical results are likely to follow. There may be less incentive to recycle plutonium as fuel and to undertake the associated developmental work necessary to do so. Also, operation of certain types of reactors and their fuel cycles will probably be modulated so as to benefit by producing plutonium for AEC—doubtless in a manner that makes for less efficient production of power and use of fissionable material. There are many, many other illustrations one might give, but perhaps these are sufficient to demonstrate the point that discretionary price practice helps determine reactor technology.

Turning now to the second question, let us try to weigh the uncertainties of discretionary pricing against the uncertainties of fuel cost estimates introduced by possible changes in future technology. We have already seen how most of the total fuel cycle cost is influenced by discretionary pricing. Similarly, components of fuel cost may vary widely as a result of technical uncertainties that experience alone can diminish. Some of these technical, and statistical, uncertainties in the fuel cycle have been recently illuminated by the *Survey of Initial Fuel Costs of Large U. S. Nuclear Power Stations,* a major study prepared in December 1958 for the Edison Electric Institute.

This report will help us to illustrate the sometimes opposing and sometimes complementary effects of possible changes in price practice and in technology. For example, in the fuel cost component described as "fuel fabrication," now the largest single component of total fuel cycle cost, AEC price policy has no direct influence on cost. This service is performed in private facilities under conditions of competitive price. The uncertainties in fabrication cost stem chiefly from changes in technology and the economies of scale. These are sufficient to require an uncertainty figure of up to ± 35 per cent of estimated fuel fabrication costs, even for techniques immediately in prospect.

But let us take a component of the fuel cycle where the uncertainties of price policy and of technical advances are both directly involved, and in a complementary manner. This is vividly seen in the influence of shorter or longer burnup on total fuel costs. Long irradiation is one of the root problems in reactor technology and costs. As the most important factor influencing fuel cost, it automatically affects all the other components, and in addition determines the direct cost of the fuel consumed. The cost of the U-235 consumed is derived from the price schedules established by AEC for enriched uranium— the higher the enrichment the higher the cost. Now, the effect of longer burnup on total fuel costs is the same as that from reduced prices for U-235. Because longer burnup is confidently expected, we must ask whether the U-235 prices set by AEC at the beginning of 1955 are more likely to fall, rise, or remain the same.

The cost of the original natural uranium fed into the gaseous diffusion plants has dropped appreciably. The average AEC prices paid for domestic production of U_3O_8 decreased from $11.94 per pound in fiscal year 1956 to $9.24 per pound in 1959. There is little likelihood that these costs will rise. Dr. Libby indicates in his paper that the AEC has more uranium in prospect than it seems to know what to do with. Because the other cost elements of enriched uranium have very probably not increased since then, U-235 prices might in due course be significantly reduced for this reason alone. We can prudently assume, then, that fuel technology and U-235 pricing practice will be mutually reinforcing in assuring future reductions in total fuel costs.

Just the reverse seems true of the 4 per cent use charge. Although longer burnup is certain to reduce the required inventory of fuel, the uncertainties of the fixed fuel charge are all in the direction of an increase, if any change at all is to occur. By way of analogy, private electric utilities customarily assign a 12 per cent carrying charge to coal stocks.

We have not succeeded fully in pinning down the relative importance of cost uncertainties deriving from changes in technology as distinguished from possible changes in price practice. Surely we can say that both are significant for the future direction and composition of fuel costs.

We come now to the third and fundamental question, the rationale underlying government price setting. The stated policy of AEC, as given in *Atomic Energy Facts,* published by the Commission in 1958, is this:

> In general, it is the policy of the Commission to supply the materials and services needed by industry only to the extent that they are unavailable commercially. Wherever practicable, the Government intends to reduce or eliminate its sales and services as industrial sources become available. Prices and charges are based upon the principle of the recovery of full costs and include direct and indirect expenses (including depreciation) plus an added factor to cover overhead, interest on investment, process improvement, and expenses not subject to absolute determination.

In more direct language this means AEC very much wants to "get out" and to turn the business over to industry. Where it cannot, however, the AEC will still charge prices that are sufficiently high as to assure the Controller General no public property is being disposed of at a discount—or loss to the government. This is prudent administrative policy, but the reference to full recovery of costs compels the question: What costs, and how are they to be defined? Suppose, for example, the AEC facility involved—such as heavy water production—is surplus and would be shut down if the civilian demand were not in prospect. Should AEC charge only out-of-pocket expenses, or should it assign depreciation and other fixed costs?

Or take a facility, such as the huge gaseous diffusion plants, some of which have been producing enriched uranium largely for weapon purposes for as long as fifteen years. Should rather small diversions from plant capacity for civilian requirements be accounted for in the same cost fashion as weapons product? Furthermore, production of very large amounts of enriched fuel for nuclear power reactors might well require a different manner of plant operation, emphasizing low rather than high enrichment and requiring perhaps only a part of present plant capability or uranium supply. Might this prompt a different approach to civilian prices? Or, again, should average or marginal costs be the benchmark for the civilian price schedule? These speculations may sound irritatingly academic. They are not at all academic, as AEC staff members well know. Decisions already taken on such choices affect significantly both the present and future economics of nuclear power.

Another problem confronting a conscientious price administrator, such as AEC, is whether to avoid any changes in price, and thus avoid unsettling the already-difficult engineering estimates, or to make occasional changes as dictated by well-established trends in cost experience? This may soon be a serious choice in pricing enriched uranium. Is the price schedule, set several years ago, in line with present experience and reasonable prospects for all the gaseous diffusion plants? New plant performance, possible changes in plant depreciation, in maintenance requirements, in plant efficiency, and in properly assignable AEC overhead, would all seem to suggest an overdue *downward* adjustment—quite independent of the lower natural uranium prices already mentioned. Yet, no change has occurred. Hence, AEC seems to rank high the desirability of sticking with a price once it is set and accepted.

The price of U-235 is the very heart of the price structure for nuclear fuel and associated services. Any sharp reduction would ramify through all reactor designs and fuel systems. Here are some illustrations of the derived effects as seen in alternative fuel systems. Plutonium and U-233 are priced largely on their theoretical heat values as potential future substitutes for U-235, so their fuel values would drop correspondingly. This might further encourage the use of these materials as reactor fuels and discourage their production for sale back to the government. Another example: There is a striking contrast between the present relatively low price of natural uranium, on the one hand, and the much higher price of *very* slightly enriched uranium on the other. This relationship is crucial to the comparative advantage of reactors fueled with slightly enriched uranium as against reactors using natural uranium. Were this kink in the price schedule to be made less marked, then enriched uranium reactors, which already seem to offer lower total generating costs, would become much more convincingly the preferable reactor route to follow, abroad as well as at home. The present irregularity in the price schedule obviously raises interesting possibilities of blending different enrichments to secure lowest net fuel cost.

What can these facts and trends tell us about future pricing practices and policies? *First,* they suggest that AEC's price structure—built on the costs of U-235 and on full-recovery-of-costs theory—will be subjected sooner rather than later to pointed questions by the industry that is now trying so painfully, and with less than startling success, to bring competitive nuclear power home to the United States. A leading reactor manufacturer, who is also an AEC contractor and hence familiar with actual costs, recently suggested a re-examination of AEC prices. Quite significantly, he did not include the 4 per cent use charge among the prices that should be reconsidered. Present price policies and practices—as seen in the dubious 4 per cent use charge, the obsolescent price schedules for enriched uranium, the incongruity of plutonium prices, and the hypothetical, imputed costs for chemical processing services—all seem to call for *unified* re-examination before this new industry moves into the next stage of development. Further distortion in the technology and unnecessary pricing uncertainties might be avoided, leading to less wasteful allocation of private and public resources.

Second, the price policies set down in the 1954 Act are broad, obscure, and in certain respects mutually conflicting. For example, Section 53-d on the use of fissionable material, specifies a "reasonable charge," but is qualified by five other considerations; Section 56 specifies a "fair price" for producing fissionable material, but with two major qualifications; Section 161-m instructs AEC to establish prices on materials or services furnished by AEC that "will provide reasonable compensation to the Government . . . and will not discourage the development of sources of supply independent of the Commission." Thus the pricing policies of the Act and the legislative history may well be assigned some responsibility for apparently divergent pricing practices of AEC. Granting that appearances can be deceiving, it would seem from a public point of view that the life of AEC as price administrator has been torn by conflicting demands. As in many other aspects of the Act and particularly reactor development, AEC is compelled to act in a multiple role—as producer, buyer, seller, promoter, trader, as well as accountant. One cannot help but sympathize with AEC's dilemma, recognizing that the agency is at one time functioning as a government procurement officer (plutonium), as banker and custodian of government property (low-interest loans of U-235), as a benevolent monopolist (U-235 and heavy water), and as a harried accountant (chemical processing charges).

Fortunately, the economic thrust of these divergent roles proves in practice to be partially compensatory. The *net* effect on total fuel costs and on technology is significant, but not overwhelming or irreversible.

Furthermore, and perhaps most important, international competition in nuclear power development can probably be relied upon to help drive prices and costs down and reduce the present remarkable range of uncertainty. This competition is appearing in a variety of forms:

In the competition between natural uranium and enriched uranium reactors.

In the possibility of Western Europe building some gaseous diffusion capacity, uneconomic though this would be.

In the ever-present technical competition of the large, diversified nuclear power program of the U.S.S.R. revealed at the Second Geneva Conference, 1958.

E. BLYTHE STASON

Human Resources in an Atomic Age

DR. LIBBY invokes his wide experience to discuss the reserves of raw materials available for the production of nuclear energy. He gives us the benefit of reassuring conclusions. He notes that we may find ourselves short of uranium if we depend, in the production of electric power, solely upon the burning of U-235; but he observes that we are moving forward rapidly toward better utilization of our uranium supplies by converting the relatively common U-238 into plutonium-239 by making use of breeder reactors such as the 100,000-kw Fermi reactor now under construction at Monroe, Michigan. He also calls atten-

EDWIN BLYTHE STASON, dean of the University of Michigan Law School since 1939, started his professional career as an engineer. From 1919 to 1922 he was assistant professor of electrical engineering, University of Michigan. After two years of practicing law in Sioux City, Iowa, he became professor of law, University of Michigan, in 1924. From 1938 to 1944 he was provost of the University. Among many special assignments, he served on the Hoover Commission Task Force on Legal Services and Practices, 1953-54. In 1958 he was chairman of the Michigan Governor's Study Commission on Tax Administration. Since 1955 he has been managing director of the Fund for Peaceful Atomic Development, and has written extensively in the atomic energy field. Mr. Stason was born in Sioux City, Iowa, in 1891, and received his A.B. from the University of Wisconsin, B.S. from Massachusetts Institute of Technology, and J.D. from the University of Michigan.

tion to the large supplies of thorium, heavy water, and lithium, all of which are available in quantity. Accordingly he quite reasonably concludes that the supply of nuclear fuels will meet the needs of the future. In addition Dr. Libby points out dramatic possibilities in the utilization of atomic explosions for large-scale peaceful operations, including even the production of electric power by the containment and use of the heat energy that is generated at the point of detonation. With all of these possibilities in view we may look ahead with confidence that our natural resources will be up to the task of meeting anticipated needs.

Dr. Libby's paper suggests two types of unique problems that will confront those who are involved in the atomic future. First, his discussion raises certain legal problems of a rather novel character not hitherto discussed to any great extent. Then, second, his treatment of natural resources also suggests exploration of the human resources requisite to the development of peaceful uses of atomic energy.

Legal Problems

We have in recent years become reasonably familiar with certain legal problems connected with atomic energy—for example, tort liability for radiation damage. We know that overexposure to radiation can result in cancer, leukemia, genetic damage, shortening of the life span, and possibly other human ailments. We also know that the presence of radioactivity may render property useless, crops inedible, and may even require the evacuation of homes, farms, and other valuable property. We know that these consequences may not only follow the rare catastrophic accidents, but they may be involved, although to a less dramatic degree, in the more numerous minor incidents, or they may even be caused by radioactive discharges to the environment resulting from more or less "normal" atomic operations.

All of these matters present new challenges to the law.

Among other features, we are challenged by certain difficulties in proof of causation and new aspects in proof of damage. Again, the statutes of limitations, which are now limited to short periods of one to three years in order to assure the trial of cases while witnesses are still living and of good memory, must be amended to take account of the contingencies which will arise from the long delay between actual over-exposure and the manifestation of damage. All of these problems have been discussed in recent literature; and although they are all brought to mind by Dr. Libby's paper, space does not permit more than this brief suggestion of their presence.

However, Operation Plowshare is another matter—something quite new. We are to use atomic explosions to shatter or move large amounts of earth material, or to facilitate the recovery of oil from the ground. The tremendous heat energy generated by atomic explosions will be trapped and retained within the earth to heat water. The steam thus produced will be used to turn turbines to produce electric power, and for other interesting purposes. Or the tremendous number of neutrons generated in the explosions will be captured and directed to useful purposes.

But what new legal problems will arise from Operation Plowshare? Consider, for example, the authority of the Atomic Energy Commission to fix charges under the 1954 Act. How shall it fix charges for producing atomic explosions, in view of the windfalls thereby made available to oil companies whose oil production will be greatly enhanced? Or, in another setting, how shall it fix charges for dredging a harbor by atomic explosion? Here there would be no windfall, but instead the public harbor authority would presumably be able to render better service for the community. Shall an atomic explosion for such purpose be staged at a reduced price? Shall there be a graduated price schedule, depending upon the size of the windfall? Statutory amendments to cover the point are clearly indicated.

There are some other questions. Should the Commission, or can it under present statutory authority, engage in what is in effect a commercial service for those who wish to expedite the recovery of natural resources from the earth? If such power exists, how is the Commission to choose between the many companies that will knock at its doors, asking for explosions to produce windfalls? Questions of priorities, as well as charges, must be given attention, and again, appropriate amendments to the Atomic Energy Act are indicated. It is certain that the statutory treatment of these matters must be far more sophisticated than that currently provided in the legislation.

Then again, what about possible damages to third parties inflicted by the Plowshare type of operation? The financial responsibility of the private operators requesting the atomic explosion service may or may not be adequate. It is doubtful whether sufficient private liability will be available, and the present law makes no provision for government assistance in covering risks of deliberate nuclear explosions.* Under such circumstances the public is going to take considerable interest in commercialized atomic explosions. It may be expected that, unless public confidence is developed by careful and adequate measures, there may be substantial public resistance to widespread use of the new potentialities of the atom. The public will doubtless insist upon representation at public hearings, thus to assure protection of its health and safety.

In short, while Operation Plowshare presents some fascinating technological possibilities and reveals some important economic values, I can also foresee a large new line of legal problems, making further business for members of my profession.

Human Resources

Now let me turn to another problem suggested by Dr. Libby's discussion of natural resources, namely that of the human re-

---

* Section 170 of the Atomic Energy Act, under which the federal government does supplement the maximum insurance available from private companies, applies only to damage from power reactors.

sources available to carry on atomic activities.

A year and a half ago the Atomic Industrial Forum published a thought-provoking survey of scientific and engineering manpower requirements as envisaged for private atomic industry. The survey was conducted under contract with the United States Atomic Energy Commission. Questionnaires were sent to some 3,000 firms throughout the country. It was concluded that the need for scientists and engineers for privately supported atomic energy activities would more than double during the succeeding three years.

The Forum report covered only the needs of private industry. In addition there are the ever-increasing personnel demands of the Atomic Energy Commission and its contractors, together with the requirements of the Army, Navy, and Air Force and other government agencies engaged in activity in the atomic field.

Also there are the needs of other nations, particularly those relatively underdeveloped areas that must, of necessity, rely in large measure upon the United States to furnish the technical training in atomic techniques so necessary to acquire a foothold in the nuclear future.

The total world-wide need for trained scientists and engineers undoubtedly stands as one of the principal challenges of the day. Moreover, the free world must not overlook the fact that Communist countries, under the leadership of the Soviet Union, are forging rapidly ahead, supported by an educational system well qualified to turn out the technicians needed for atomic development as well as other phases of the technological future.

With this challenge in mind, let us survey the principal facilities currently available in this country for meeting the needs, for supplying the manpower requisite to supply the increasing demands.

College and university programs. About 100 of the 150 accredited engineering colleges in the United States now offer well-developed nuclear programs. Colleges and universities have an estimated 5,300 scientists and engineers who devote at least a part of their time to teaching and research in the atomic energy field. These are the persons who will be very largely responsible for training the atomic scientists of the future and for keeping the supply of trained personnel abreast of the demand. The colleges and universities have done remarkably well in adjusting themselves to the atomic challenge and doing their share to meet it, even though funds for staff and equipment have not been easily obtained, and in the earlier years classification of information constituted a most difficult obstacle.

Atomic Energy Commission educational activities. The Atomic Energy Commission has long been aware of the need for technical personnel and has taken important steps to fill it. The following summary will reveal the extent of Commission interest in educational activities.

(1) In 1948 an association of thirty-four southern colleges under Atomic Energy Commission contract established the Oak Ridge Institute of Nuclear Studies for the purpose of providing training in the safe and efficient use of radioisotopes. Provision was made for short courses of about four weeks each. Each class includes about thirty persons, and altogether some 2,800 scientists and engineers from every state in the Union and from some thirty-seven foreign countries have completed the courses.

(2) To meet a serious need for advanced training in reactor technology, the Oak Ridge School of Reactor Technology was established in 1950. It has performed an essential service in training individuals in the specialized field of reactor technology. Originally most of the basic scientific data were classified and "Q" clearances were required, but at the present time declassification has proceeded to such an extent that the field is almost completely wide open. Over 500 scientists and engineers have been trained in this program.

(3) In 1955 at Argonne National Laboratory the Atomic Energy Commission established The International School of Nuclear Science and Engineering, designed primarily, although not exclusively, to make available to persons from other countries graduate level instruction in nuclear science and engineer-

ing. By accepting also a few students sponsored by American industry as well as those from other countries, the Commission has afforded an opportunity for friendly working relations between United States scientists and those from abroad. Altogether some 408 scientists and engineers have been trained in the International School, 378 of whom have come from forty-three different foreign countries.

(4) In September 1958, during the Second International Conference at Geneva, the Atomic Energy Commission announced the inauguration of three new courses for advanced students from friendly foreign countries. They are a twelve-month course in Reactor Hazards Evaluation, and another nine-month course in Reactor Supervision, to be given at Oak Ridge National Laboratory beginning in February 1959 and annually thereafter; also a six-week course in Radio-chemical and Counting Procedures, to commence at the New York Health and Safety Laboratory of the Commission in October 1958, and to be repeated four times a year thereafter.

(5) In order to further promote the training of scientists and engineers in the atomic field, the Atomic Energy Commission has provided financial assistance to permit educational institutions to acquire training reactors, teaching aids, technical apparatus, laboratory equipment, and nuclear materials to be used in nuclear technological courses.

As of the summer of 1958 a total of approximately $10,900,000 had been awarded to 102 educational institutions to provide the necessary laboratory equipment required to broaden nuclear science and engineering curricula, including teaching reactors and subcritical assemblies. In addition, a total of upwards of $1,200,000 had been made available to eighty-seven schools to assist in equipping college and university laboratories for training in life sciences as affected by nuclear activities. Also the Commission has made many loans to educational institutions of natural uranium, uranium enriched in the isotope 235, and neutron sources.

Any appraisal of the Commission's educational program must accord first-class commendation for the broad and imaginative service which has been and is being rendered on many educational fronts and with full appreciation of the serious importance of keeping up with the human resource needs.

Training of high school and college teachers in nuclear science. Admiral H. G. Rickover has pointed out that the supply of prospective teachers for secondary schools in scientific fields, already inadequate, has been dropping steadily. For example, he says college graduates prepared to teach mathematics have decreased from about 5,000 in 1950 to about 3,000 in 1958. Corresponding figures for teachers of science show a reduction from about 9,000 in 1950 to about 5,000 in 1958. Moreover, many of those who are qualified to teach stay away from the teaching profession primarily because the fiscal rewards are greater in commerce and industry. A somewhat similar shortage of faculty personnel with experience in nuclear engineering has been felt by the colleges and universities.

The Atomic Energy Commission, conscious of these shortages, offers summer institutes of two months each conducted at Argonne National Laboratory, Brookhaven, Oak Ridge, and elsewhere, to give college teachers advanced instruction in reactor physics, metallurgy, chemical separation, reactor instrumentation and controls, and other pertinent subjects. For these summer courses the Commission provides 50 per cent of the cost of living for each participant up to $750, matching amounts being provided by the participants' respective academic institutions. For secondary school teachers the Commission, jointly with the National Science Foundation, maintains a dozen or more eight-week summer institutes in radiation biology with all expenses paid. The Commission deserves great credit for its efforts to meet the educational challenge posed by atomic development.

The United States Navy training program. It has been reported that the United States Navy trains more than 2,000 young men each year to serve as nuclear technicians to man atomic submarines and future atomic surface ships. The Navy wishes to develop a pool of technically trained persons upon whom it can draw to meet the expected needs of the future.

Industry training programs. Industry is taking a hand in training its own personnel for the handling of radioactive materials. The General Motors Company, for example, has its company school to train key workers in handling such materials. It offers a ten-week program designed to overcome the shortage of trained personnel available for the company's radiation laboratories.

The General Electric Company's partial answer to the possible shortage of atomic scientists and engineers is a special school at Hanford, Washington, attended by personnel working in the Hanford atomic complex. The curriculum includes not only the handling of radioactive materials but also reactor design. Arrangements are made for co-operative studies at the University of Washington, Washington State College, the University of Idaho, and Oregon State College.

From the foregoing summary of the principal educational activities in this country we can conclude that there is widespread recognition of the serious need of developing the human resources for atomic activities, and much is being done to bridge the gap. There is room for guarded optimism on the educational front.

Fellowships and scholarships. Many highly qualified young men and women are quite unable to finance the costs of higher education, especially of advanced graduate-level education essential for adequate training in nuclear science and engineering. Financial assistance must be forthcoming in generous quantities if the personnel challenge is to be adequately met. Here again we note some first-class results.

Many of those who have attended the government schools, as well as colleges and universities, have been sponsored by the industries in which they have been employed. Moreover, the Atomic Energy Commission grants upwards of 150 special fellowships each year in nuclear energy technology, about seventy-five for radiological physics, about eight for industrial hygiene, and a few in industrial medicine. As previously mentioned, the Commission has also matched like contributions from educational institutions to support approximately 150 university faculty members attending summer courses and institutes.

The International Cooperation Administration has provided financial support from its Mutual Security Funds for a substantial number of advanced students from foreign countries who come to the United States to study in the International School of Nuclear Studies at Argonne National Laboratory. And many foreign students have been financed by their respective governments.

A new and important organization, the International Atomic Energy Agency, is making a very substantial contribution to the available supply of nuclear fellowships. In its program and budget for 1958-1959 the Agency announces the availability of $250,000 from its general fund to provide fifty fellowships at $5,000 per year each. The duration of the fellowships financed by IAEA general funds is limited to two years. In addition, a number of the member states of the agency have established about 140 additional fellowships, supporting study in their respective national institutions. It is understood that an even larger number will be available in 1959-60.

There is one rather significant point to be observed by those who are concerned about the position of the free world vis-à-vis the Iron Curtain countries in the atomic race. Among the 140 fellowships offered by member states for training of foreign students in national institutions, the Union of Soviet Socialist Republics announces that it will make available forty-five for study within the Soviet Union. Poland offers five, and Rumania nine. The United States of America, with all of its wealth of educational opportunity, offers only twenty. The Soviet Union is seemingly more concerned than we are over the establishment of friendly relationships with scientists in other countries. Young scientists who are invited to the Soviet Union for advanced study in the nuclear field will no doubt be well treated, and when they return to their respective countries they will serve as foci of friendship and good will toward the U.S.S.R. We could well afford greater liberality to promote such foci of interest for the United States.

So far as domestic development of laboratories and teaching facilities are concerned, we are well on our way toward the utilization of the available human resources for the atomic future and for the training of new personnel to fill the anticipated needs. The same is true of domestic fellowships. Those in positions of responsibility can properly claim credit for good performance. A careful look at the entire scene, however, reveals certain shortcomings. I mention three.

First, I am concerned, as are so many others, over the inadequacies of our secondary school programs, which are falling short in arousing student interest and in equipping young men and women to carry on the technical tasks of the scientific future. Our treatment of mathematics, physics, chemistry, and the biological sciences during the high school years is deplorable. Not only are the essential courses lacking in the curricula of many schools, but the teaching staff is clearly inadequate to the task posed by the technological world of tomorrow.

The Soviets in their ten-year college preparatory program accomplish far more than we do in our twelve-year program in the way of building an educational foundation for participation in atomic and other scientific activities. They teach surprisingly large amounts of mathematics, physics, chemistry, and biology, and the youngsters work hard. In this respect we in the United States are falling short of the best utilization of our human resources. Yet I have faith in the potentials of our free-wheeling system. We can realize them through better discipline and wiser selection of subject matter.

Second, I am concerned over the fact that we are not exerting an all-out effort to provide more fellowship assistance for the bright and eager young men and women in other portions of the world, particularly but not exclusively the underdeveloped areas and the so-called "uncommitted areas," where such assistance is sought and where it would prove of great value, not only to the people themselves but to the promotion of good will toward the United States. Moreover, I wish we could help certain hard-up universities in far places to establish laboratories and training centers so they could help themselves to the atomic education they need and greatly desire.

The world's hopes, aroused in 1953 by President Eisenhower's "Atoms for Peace" talk before the General Assembly of the United Nations, have neither been quenched nor satisfied. The Soviets will bridge the gap if we don't. For the sake of future relationships between the United States and the peoples of other lands, there should be a concerted effort to extend the possibilities and facilities for the training of nuclear scientists to help other nations achieve the benefits of atomic science. Not only government but also private industry could well afford to lend a hand. The effort would pay excellent dividends, not only in good will, but also in foreign trade.

Third, what about the impact of atomic development and, indeed, of all large-scale contemporary and future scientific and engineering progress upon the human beings who participate in it? The scientist or engineer in an atomic activity is, in most instances, a human cog in a huge machine, a part of a great team. He and his associates are dealing with immense masses of equipment, extraordinarily valuable materials, extremely complex and hazardous processes, all a part of a huge capital investment in a multi-million dollar enterprise, whether it be publicly or privately owned. There is little if any "small business" in atomic energy. Nuclear scientists and engineers are not independent craftsmen or artisans, or even lone wolf laboratory or research technicians. They are not their own masters.

We are not speaking of just a few members of society. The 35,000 atomic specialists of today will increase tenfold when atomic electric power becomes economically competitive and when the use of radioisotopes and radiation becomes really widespread. Many more technically trained persons will be involved in other new enterprises—jets, missiles, space age equipment, etc.—all large, complex, expensive activities. A very large portion of the intellectual aristocracy of the nation will be engaged in highly regimented enterprises.

This is all a part of the technological future. What will be its effect upon the freedom of mind and action that has so characteristically been a part of America? In some countries the individual counts for nothing; the state is all. We in the United States, on the contrary, believe in the importance and dignity of the individual. What will become of this dignity in the future, when hundreds of thousands of our finest human resources are regimented by the technological managers? We may assume that the regimentation will be all for the common weal. But regimentation it will be, whether it be exercised by bureaucrats in governmental offices or directors of huge corporate enterprises. Taking the long-range view, we must, as a part of the evolutionary process, anticipate the need of adjusting ourselves more and more to a regimented human environment for most of those involved in the atomic and technological future.

vi

NEW KNOWLEDGE
FROM OUTER SPACE

Lee A. DuBridge:

SPACE EXPLORATION: HOW AND WHY?

Alan L. Dean:

MOUNTING A NATIONAL SPACE PROGRAM

Philip C. Jessup:

THE INTERNATIONAL OPPORTUNITY

LEE A. DuBRIDGE

Space Exploration: How and Why?

IS OUTER SPACE a resource? If it is (and after all it has been chosen as the final subject of this Resources for the Future Forum series), one very positive thing can be said about it immediately: there is plenty of it! In fact, the very vastness of outer space is itself so extraordinary and so incomprehensible that I shall return later to a discussion of it.

Outer space is not only plentiful; it is also durable. It never gets used up; it is certainly not a dwindling resource. In fact, if you want to speak precisely, the quantity of outer space is

LEE A. DuBRIDGE, physicist, has been president of California Institute of Technology since 1946. Some years earlier, 1926-28, he was associated with Caltech as a National Research Council fellow. He was assistant professor of physics, Washington University, St. Louis, 1928-33, associate professor 1933-34; professor of physics and chairman of the department, University of Rochester, 1934-46, dean of faculty of arts and sciences, 1938-42; investigator, National Defense Research Committee, and director of the Radiation Laboratory, Massachusetts Institute of Technology, 1940-45. He was a member of the General Advisory Committee of the Atomic Energy Commission from 1946 to 1952; Naval Research Advisory Committee, 1945-52; Air Force Scientific Advisory Board, 1945-49; member, President's Communications Policy Board, 1950-51; member, National Science Board, 1950-52 and 1958-; chairman, Scientific Advisory Committee, Office of Defense Mobilization, 1952-56; and a member of the National Manpower Council from 1951 to the present time. He is a trustee of the Rockefeller Foundation, of The RAND Corporation, and

rapidly increasing! Because of the expansion of the universe, the radius of space is increasing at a rate nearly equal to the velocity of light. This adds quite a lot to the volume of space every year.

A resident of New York, Chicago, or Los Angeles must certainly regard space as a pretty transient resource, as he sees the space available to him dwindling each year at a rapid rate. Naturally, therefore, he looks to *outer* space in the hope that most of his neighbors may some day be transported out there. On this point we cannot offer our harassed city dweller much hope. After all, he or his neighbors could, if they chose, move at any time to Texas or Alaska, to the Mojave Desert or to Death Valley, or to many other places. If he has doubts about the living conditions at any of these places, he should contemplate for a moment the living conditions on the moon! If he does not like the desert because of the scarcity of water and food, why would he choose the moon where there is also not even any air?

The entire surface area of the moon is only one-sixteenth of the surface area of the earth, or one-fourth of the land area. The whole surface of Mars has an area about equal to the land area of the earth. Hence, if we are looking for extra space to which to transport an excess population, it would clearly be cheaper to build a colossal floating platform over the surface of all the earth's oceans. This would multiply our living area by four, whereas the moon and Mars combined would provide us less than a factor of two. Furthermore, I repeat, the earth has air— blessed air!

In fact, if it is space to accommodate a too rapidly growing population that we are concerned with, I refer you to Dr.

---

of the Mellon Institute for Industrial Research. He received the King's Medal for Service (Britain), 1946, the Research Corporation Award in 1947, and the Medal for Merit (U.S.) in 1948. Mr. DuBridge was born in Terre Haute, Indiana, in 1901, and received his A.B. degree in 1922 from Cornell College (Iowa), and his Ph.D. in 1926 from the University of Wisconsin.

Beadle's paper that begins this book in which he discusses the problem of population control. That, I assure you, is a much more manageable solution to the "space" problem than colonizing other worlds.

What, then, is the connection between outer space and "resources for the future"? To tell you the truth, it seems pretty likely that for the next few years the exploration of outer space is likely to turn out to be one of our best methods of *using up* natural resources rather than conserving them or increasing them. The Bureau of the Budget, I predict, is not going to like outer space. Any budget item that amounts to a billion dollars or more a year is bound to be a real headache for the budget-makers. And I predict they will ask some pointed questions about where all that money is going. Indeed, I hope they will. For that is my money and yours. Of course, a billion dollars is only $6 apiece for each of us, on the average, but still I'd rather keep that $6 than see it wasted. Furthermore, that $6 represents an expenditure of natural resources. A lot of steel, copper, oil, coal, many other valuable materials, and much human labor can be bought for a billion dollars, and it would be a good thing for the American people to try to understand what the investment is for and what returns it is likely to yield.

It is frequently suggested that on the moon or Mars, or some other planet, we may find huge stores of valuable minerals— gold, copper, uranium, or something else. I can guarantee you won't find coal or oil on the moon, for these come from *living* things! But I think it is very clear that it would be far cheaper to extract gold from sea water or uranium from granitic rocks than to haul them from the moon. For we are really not running out of these minerals here on earth; we are only running out of cheap sources of them. The moon or Mars can hardly be regarded as cheap sources for anything.

Let me hasten to make it clear that I think a good sound program of space research, space exploration, and possibly space exploitation, *is* worth a billion dollars a year to us— possibly very much more than that. I favor a bold, imaginative, and extensive program of space activities covering both military and civilian possibilities—including many research ventures whose potential value, whether military or civilian, cannot possibly be foreseen. My only hope is that this program can be based on realities rather than on fancies. I do not subscribe to the thesis that to be bold one must go off half-cocked. Quite the contrary, the boldest programs are just the ones that require most careful planning. Furthermore, this planning must stick close to realities. It must be done by scientists and engineers whose vision is backed by solid facts, and not by space cadets whose only source of information is science fiction or the comic strip.

It will be my purpose in this paper, therefore, to examine space activities from the point of view that the greatest resource to be gained from them is knowledge—new knowledge about our own earth, as well as about outer space; and new knowledge about the techniques of getting out there to gain more knowledge. After all, no human resource is more valuable than knowledge. And when we contemplate what a vast sea of ignorance we face in outer space, it is natural that we should be impatient to get on with the task of replacing ignorance by knowledge. I shall discuss, first, certain matters related to the nature of the space environment; second, some of the goals to be sought in space exploration; and, third, certain of the technological problems we face in attaining these goals.

The Space Environment. At first thought it might seem that "empty space" is something about which there is not very much to say except that it is empty and big. Closer examination, however, shows that while space is certainly big it is not empty, and it will be instructive to review some of the things we know about it.

First, the bigness. It is really meaningless to talk about the size of space itself, but it is not meaningless to talk about the distances between the various tangible objects in space. In fact, some of these distances are so enormous that it pays to take a look at them before we talk too blithely about the journeys we are going to take out to this object or that. It is not very useful to express these vast distances in miles, because the numbers are too huge to carry meaning. We could follow the lead of the astronomer and express them in light years—that is, describe the distances in terms of the time it takes a beam of light to traverse them at a speed of 186,000 miles per second. This, however, gives an inadequate impression of the distances because light travels at a speed thousands of times greater than that which we can hope to give any material object in the foreseeable future.

I shall, therefore, express these distances in terms of the time required for a possible space vehicle to traverse them. I shall arbitrarily assume that we have a space vehicle which can travel at a speed of 25 miles per second, or 90,000 miles per hour. This is three and one-half times the speed of escape from the earth; it is just about equal to the speed required to escape the sun's pull when in an earth-like orbit; it is also 50 per cent greater than the earth's orbital speed around the sun. We remember, of course, that no actual space object projected on its journey will ever maintain a constant speed—its speed will rise or fall as it approaches or recedes from bodies which exert gravitational attraction. Suppose, however, that some space vehicle could be created which could go at the constant speed of 25 miles per second in any arbitrary direction, being speeded up or slowed down as necessary to compensate for changes in gravitational potential. How long would it take this vehicle to travel from the earth to various points in space? Here are a few sample items:

To go to	The time required is
The Moon	2.9 hours
Mars (nearest approach)	16.0 days
The Sun	43.0 days
Uranus	780.0 days
Pluto	4.5 years
Alpha Centauri (nearest star)	30,000 years
Center of Milky Way	560,000,000 years
Andromeda Nebula (nearest spiral galaxy)	15,000,000,000 years

The conclusion is obvious: All points *within* the solar system (the first five items above) are well within reach of our imaginary vehicle in times reasonable compared to a human lifetime. However, *no* known object *outside* our solar system comes within a factor of a thousand of being accessible. It is true we can some day probably exceed the speed of 25 miles per second. But the 25 *thousand* miles per second required to bring the nearest star within reach is not in sight. In brief terms: inter*planetary* but not inter*stellar* space is now open to conquest.

Probably the most conspicuous property of inter*planetary* space is the existence of the all-pervasive gravitational field. The intensity of the field fluctuates greatly, depending on one's position relative to the sun or one of the planets. For example, a body which weighs 100 pounds on the earth would weigh only 25 pounds at 4,000 miles from the earth's surface, and 1 pound at 36,000 miles (40,000 miles from the earth's center). It will weigh 16 pounds on the moon, 38 pounds on Mars. If we recede from the earth, but still remain at the same distance from the sun, the latter's attraction with a force of 1/10 of a pound will eventually predominate. This force, in turn, will vary inversely as the square of the distance to the center of the sun and will have appreciable values out to distances of billions of miles.

A gravitational force inevitably means an acceleration and no object in the solar system can remain at rest. Hence, any object projected from the earth, if it does not return to earth, will go into some sort of orbit about the earth; or if it escapes the earth's pull, into an orbit around the sun. It would not, without a further "push," go into an orbit about any other object. These closed orbits are always ellipses (a circle being a special and rather improbable form of an ellipse). If an object's velocity is high enough for it to escape altogether from the attracting center, the path will be a parabola or hyperbola. Furthermore, once the propulsive force has ceased to act on the object (when, for example, the rocket has burned out), then the precise path of that object in the gravitational field is determined (and its velocity determined too) for all time to come until another propulsive force is applied, such as another rocket impulse or the retarding effect of friction if the object enters an atmosphere. Thus, a satellite projected into an elliptical orbit around the earth at an altitude sufficiently great to avoid atmospheric friction will continue in a predictable orbit forever. The particular orbit to be followed will, furthermore, be determined solely by the position and the direction and magnitude of the velocity at the instant the propulsive force ceases. Two objects starting from different initial positions, or with different initial velocities, cannot attain the same orbit. Nor can two objects traverse the same orbit with different speeds; an orbit is not a race track in which one vehicle can overtake another. Nor can two objects in different non-intersecting orbits ever have the same speed. The object farther away from the attracting center must always be going more slowly.

I emphasize this point of the inevitability of motion and the predetermination of motion in a gravitational field because many discussions of space travel seem to assume that a platform can be established which can float lazily around in space like a boat on a quiet lake. Not so. A boat in a whirlpool is a more accurate analogy—it simply can't stop!

There is one particular type of orbit about the earth which presents interesting possibilities. If a satellite is projected into an orbit above the earth's equator at a height of about 22,000 miles, its period of rotation about the earth will be just 24 hours. Since the earth is rotating under it at the same rate, the object then will appear to be stationary in the sky—a moon that never sets. Actually it will be moving at a speed of over 6,000 miles per hour. This, however, is the nearest approach to a "stationary platform" that it is possible to achieve. Note also that it must be in an equatorial orbit which is very close to circular and at exactly the right height, or else it will not stay in the same spot in the sky.

Certain proposals for the use of so-called space platforms require that they remain "stationary" in the sense of hovering above the same spot on earth. Other space stations, such as those proposed as assembly and take-off points for deep-space expeditions, need not be in 24-hour equatorial orbits at all, but can be in any convenient orbit above the earth. They should, however, avoid the Van Allen layers!

The rotational nature of this "perpetual motion" gives rise to some odd results. If an object in an orbit around the earth is suddenly given an acceleration (e.g., by a rocket) in the direction of motion, it will not thereby proceed faster in the same orbit. Instead, the larger centrifugal force will cause it to move off tangentially into a new orbit farther from the earth. But, as it moves against the gravitational attraction, it will also slow down and traverse the new orbit at a slower average speed and, of course, a longer period of rotation. Conversely, a retro-rocket would cause the object to move inward and attain a higher speed. It is amusing to speculate on the many problems encountered in an environment where one must slow down in order to go faster!

A third characteristic of space is the radiation one finds in it. We know about some of the types of radiation traversing space because they can penetrate both the earth's atmosphere and the earth's magnetic field and reach our instruments. But there are other radiations which cannot reach the earth's surface and which we cannot know about until we begin serious space exploration.

There are, of course, two general types of radiation: (1) *electromagnetic waves* of widely varying length from the long radio waves through infrared, visible light, ultraviolet light, to x-rays and gamma rays; and (2) *charged particles*—electrons, protons and the nuclei of other atoms—with a wide range of kinetic energies from a few electron-volts up to possibly a billion billion electron-volts. Of all these radiations only certain radio wavelengths and the visible portion of the electromagnetic spectrum can penetrate our blanket of air, and only the more energetic charged particles can get through the magnetic field and strike the upper atmosphere. Yet, up to 1957, *all* of our knowledge of outer space had come through a study of the radiation which does get down to our instruments. Though some instruments have been sent in rockets or balloons to nearly the "top" of the atmosphere, a whole unknown universe may be revealed as we get clear above the atmosphere and away from the earth's magnetic field.

Indeed, our very first satellites to reach the regions a few hundred miles above the earth detected a new belt of radiation—the Van Allen layer—whose existence had been previously unsuspected. It consists of two clouds of high-energy electrons or protons whose origin is unknown but which appear to be trapped in the earth's magnetic field at distances from a few hundred to 12,000 miles or so above the surface. The intensity is surprisingly great—more than a thousand times the intensity of the known cosmic rays which have been measured by balloon-borne instruments high in our atmosphere. They are intense enough to be a potential hazard to human beings who might like to travel in manned satellites above the earth. And they could ruin photographic plates sent aloft to take pictures of the earth. Many a dream about space exploration has already been abandoned or modified by this discovery—and what additional unsuspected radiation streams are yet to be found no one can tell. Radio, infrared and ultraviolet telescopes as well as Geiger counters and other detectors should certainly be sent aloft as soon as possible to begin this fascinating era of discovery, which may well last for many decades before adequate knowledge is obtained.

What will be the results of all this? I do not know. New knowledge, certainly. Many surprises, probably.* Revolutionary new discoveries, possibly. But in these vast unknown radiation fields of space there certainly lie hidden many secrets about the nature and size and composition of the universe. The cosmic rays which manage to penetrate to the earth's surface have told us many things, too, about the structure of atoms and nuclei. Those which cannot reach us may teach us even more.

Around every sizable body in space we are likely to find both

* The Argus experiments, formally announced March 26, 1959, showed that high altitude atomic bursts could increase, temporarily but significantly, the density of the electron cloud trapped by the earth's magnetic field.

electric and magnetic fields. We know a little about the magnetic field about the earth; very little about any electrostatic field. We know—from rather recent observations—a little about the magnetic field of the sun. It is quite weak and quite variable. Some stars are surrounded by very large fields—as can be determined from the so-called Zeeman effect, the splitting of certain lines in the spectra. Very weak, but very pervasive fields may spread throughout interplanetary space and throughout our entire galaxy. They might have profound importance in the acceleration and trapping of charged particles —cosmic rays—and even in the large-scale transfer of momentum between the planets and between stars. Only an extended series of properly instrumented flights far into interplanetary space will reveal the nature and extent of such fields.

Goals of Space Research. Many of the goals of space research programs are obvious from what has just been said. A major task will be to learn more about the nature of the space environment itself, the radiation streams which traverse it, and the electric, magnetic, and gravitational fields which pervade it. Certain other types of space ventures must indeed await the results of the initial investigations of the properties of space itself.

However, there are many more things to be done—so many that it is difficult even to classify them. First, we may consider the tasks which may be performed by vehicles placed in various orbits around the earth, and then the additional tasks for probes which are projected farther out into the solar system.

1. *Earth satellites*. For some time to come, the most important (though not necessarily the most spectacular) scientific missions will be performed by instrument-carrying vehicles projected into orbits at distances of from a few hundred to a few thousand miles from the earth's surface. In addition to examining the nature and contents of space itself, they may be used to make observations of the earth or of other bodies. In addition to their information-gathering function, they may also perform certain service functions, such as serving as radio-

relay stations, as refueling stations or service platforms, or possibly as carriers for military weapons. This paper will confine itself to the information-gathering function.

However much we may love to learn about the moon and the solar planets and the sun itself, the earth will always be the object of primary interest to human beings. So what we can learn about the earth from observation stations circling far above its surface is of prime importance.

Even a "dead" or noninstrumented satellite, if it were large enough to be visible from the surface of the earth (e.g., 100 to 300 feet in diameter) could provide us with quite a lot of information. By observing carefully the nature, shape and perturbations of its orbit, one may learn much about the exact shape of the earth itself and the distribution of mass within it. It should be remarked that the whole science of precise orbital calculations will need much further development. Astronomers have been working for generations to evolve an exact equation for the orbit of the moon. But every new satellite presents a new and difficult orbital calculation. Computing machines now make the task much easier—but the most suitable mathematical techniques must still be worked out, and much more information needs to be acquired about the exact form of the earth's gravitational field itself. The small but important perturbations in orbit caused by the field of the moon, the sun, and other planets cause unbelievable complexities. Yet, if we are going to carry on space explorations successfully, accurately, and reliably, this new branch of the science of celestial mechanics must be rapidly developed. It is all very well to know that, once launched, the path of a space vehicle is determined for all time by the gravitational fields through which it passes. But computing its path in advance, while always possible in principle, is extraordinarily difficult in practice.

If one looked at the earth from a satellite, the most obvious observation would be the cloud pattern. A single good picture from a satellite, which is, say, 300 miles high, could—if it could be promptly transported or transmitted to the earth's surface—give a view of the entire storm pattern over an area some 2,000 miles in diameter, i.e., over much of the United States. A few dozen such pictures taken about simultaneously from properly chosen points in the northern hemisphere could, pieced together, give for the first time a complete weather picture of that whole hemisphere. It would take a good deal of research to interpret such pictures and to use them in prediction, but clearly enormous contributions to the science of meteorology are in sight. The difficulties and cost of obtaining such collections of pictures continually and reliably, and getting them back to earth stations without losing too much resolution, are of course enormous. But useful information will be obtained even before ideally complete observations can be made. One good photograph each day of the North American area would be invaluable. But even that is expensive.

The charged layers of the earth's upper atmosphere which play such an important role in the transmission and reflection of radio waves will also constitute an area of intense interest. The charge density and thickness of these layers, the influences which cause the molecules to become ionized, and how these change with time and how they depend on events in the sun or other places, will cast important light on radio, television and radar transmission problems.

The strength, shape, and variations in the earth's magnetic field out to distances of 100,000 miles could occupy the attention of dozens of properly instrumented satellites. The origin of this magnetism is still a puzzle and, though the solution to the mystery may not be found in space, pertinent information certainly will be.

If we now turn our attention from the earth to other objects in space, we find a bewildering wealth of opportunities for making observations which are forever impossible when we stay buried in our blanket of air. However much we can bless this blanket for its life-giving properties, it is still a curse to the astronomer. As has already been suggested, observatories in space which can measure radio, infrared, visible, ultraviolet and x-rays will undoubtedly reveal wholly unsuspected things about the sun, the planets and the stars. There is every reason to suppose that the radiations which cannot penetrate our atmosphere may carry just as great a wealth of information as those that do, and a new era in astronomy will dawn when space observatories become possible. Unfortunately, complete space observatories will be very expensive—but even simple ones may be most useful.

These few examples will serve to prove what a gold mine of valuable knowledge may be revealed by instrumented earth satellites.

I have said nothing about manned satellites. The first man-carrying satellite will be a tremendous achievement and the first passengers will experience a tremendous thrill. The first look that human eyes have of the earth and the heavens from a space vehicle will mark a new epoch in the annals of human experience.

But the adventure and the prestige are not the only considerations. One must examine carefully what functions men can perform that instruments cannot perform as well or better, and which functions are worth the very great extra cost of carrying a human being aloft, of keeping him alive and useful, and of getting him back alive. Certainly a vast amount of data can be collected by automatic instruments without human intervention, and space research should not be delayed until perfection of passenger-carrying vehicles. Nevertheless the human being, though he is a costly and delicate instrument to carry aloft, does have many attributes possessed by no electronic equipment yet built. If intelligently used, man can be a great asset to space research, but if he just goes along for the ride he will be a costly liability. For the next few years the human being can just as well be left at home until we really need him to do the things that instruments cannot do.

2. *Deep space probes.* At the same time that earth-satellite vehicles are being used to explore the earth's vicinity, probes to reach the moon, Venus, Mars, and eventually other planets, will be launched. Whole new mines of knowledge will be opened up as we get into a position to make visual, photographic, magnetic, and gravitational measurements in the

vicinity of these bodies. We should like to send probes to sample the atmospheres of Venus and Mars and, eventually, to land and sample the surfaces too. And there are many other observations we should like to make.

We face here, however, some deep difficulties. From the rocket point of view there are no serious problems in projecting deep-space probes into suitable orbits which will pass *near* these bodies. One might even hope some day to cause an object to strike the moon. But for the most part, in the foreseeable future, our space probes will sail past their targets and out beyond their gravitational fields to become captured in an orbit about the sun. Such objects will be lost to view forever, and the only information they will yield is that which they radio back before their batteries burn out or before they get too far away for the radio transmissions to be detected. Whereas an earth-satellite might continue in a closed orbit for years and—when larger solar batteries are available—continue to provide useful information for a long time, our space probes will be one-shot affairs and, as they get millions of miles away, there will be serious difficulties in getting signals from them at all because of the very large amounts of power required.

A very great step forward will be made when we succeed in navigating a vehicle into a permanent orbit about the moon—and can equip it with solar cells large enough to enable its radio to operate for a long time at adequate power. A great wealth of information can be gleaned from such an experiment.

Nothing short of a very elaborately equipped vehicle can hope to get into an orbit about Mars or Venus because of the delicate navigational and propulsion problems. And even if this is accomplished when the planet is at the distance of closest approach, it will be only a few days or weeks before, as the planet and its new satellite increase their distance from the earth, they will be hopelessly out of range of the most powerful radio. Thus, a rather sophisticated space technology will be required to begin to obtain information from the vicinity of even these nearest planets. To land instruments on the planets, to explore the more distant planets, and to send manned expeditions to them will be even more difficult.

A Few Technological Problems. The success of the first earth satellites and of the first moon probes has led many people to suppose that it is now only a step to the most distant and complex exploratory ventures. It is true that once the first step has been taken it is dangerous to predict that additional steps will not soon follow. It is, however, pertinent to examine the nature of some of the problems yet to be solved.

Rockets of thrust of 300,000 pounds are now available and thrusts of a million pounds are in development. These will send substantial instrumented packages into earth or planetary orbits; i.e., space probes to the region of the moon, Mars, and even more distant planets. Even manned vehicles can be placed in earth orbits with sufficient equipment for a safe return—if the journey does not last too long. A package could also be landed safely on the moon—though the payload would be quite small because of the large amount of fuel and equipment which would have to be carried up to break the fall against the moon's gravitational field.

However, when one begins to talk about sending a man to the moon and getting him back alive, one quickly runs into thrust requirements of up to 10 million pounds, calling for advances in technology which are far in the future. Space "platforms" (that is, large orbiting vehicles to which the necessary equipment and fuel can be dispatched in smaller packages and then assembled) are said to be the answer, but it is not clear whether the technology of such space stations will come more quickly than that of the large rockets. And it is yet to be decided whether a man can bring back enough more knowledge to make his journey profitable.

However, the chief technical problems of space flight do not lie in the field of rocketry. They have to do with instrumentation, control, navigation, data transmission, power supplies, and with such problems as how to get satellites, or any packages ejected from them, safely back to a given destination on the earth. I shall briefly mention only two of the many complex problems; both concern energy requirements: (1) energy for propulsion over long and complex space flights, and (2) energy for operation of instruments and equipment—especially radio equipment—within a space vehicle.

1. *Propulsion.* It has already been pointed out that rocket technology will soon make available thrusts of a million pounds or more, and this is quite adequate to lift sizable vehicles into earth orbits and into deep-space trajectories. Such thrusts would fall far short, however, of transporting manned vehicles to the moon or to a planet with enough equipment and fuel for a safe return to earth. Some wholly new ideas appear to be called for. Even if one grants that space stations will allow the fantastic loads required for such operations to be lifted in small packages from the earth, there is still much room for the development of some propulsion-energy source more useful than a mixture of kerosene and liquid oxygen.

The first thought in this field, of course, is nuclear power. In fact, the space amateur blandly dismisses all difficult propulsion problems by uttering the magic words "atomic energy." But a closer look is called for.

It is certainly true that a fission reactor in a space vehicle could supply a large amount of heat for a very long time without refueling. Unfortunately, heat alone does not provide propulsion. The heat must be imparted to some substance whose molecules, thus speeded up, are then ejected from the vehicle. The simple physics of jet propulsion tells us that the momentum (mass times velocity) of the material ejected during a given time is precisely equal to the increase in the forward momentum of the propelled vehicle. Obviously, the mass to which the velocity of ejection is imparted by the reactor heat—that is, the mass of the propellent fluid—must be carried along in the vehicle as it leaves the earth. So the limitation on the propulsive effect of a nuclear reactor comes not when the reactor runs out of heat or of uranium fuel, but when the supply of propulsive fluid has been exhausted.

What shall we use for the propulsive fluid? Simple physics again tells us that, for a given total *mass* of such fluid (for a given reactor temperature), the maximum velocity, and hence the maximum momentum, will be imparted to the fluid with the lightest molecules. This means that the ideal propellant is hydrogen. The main advantage of nuclear heat over chemical burning is that hydrogen may be used as the propellant. However, the problems of packaging many tons of hydrogen for a space journey are imposing indeed. To compress it in high-pressure gas cylinders is out of the question—for the cylinders would weigh a hundred times as much as the contained hydrogen gas. One must, therefore, cool the hydrogen to $-252°C.$, where it becomes a liquid, and keep it at that temperature until it is used up. This is a perfectly possible operation, but presents some obvious problems of storage and insulation. Furthermore, even in the liquid state hydrogen is not a very dense substance, and hence 100 tons or so of it will occupy a lot of precious space and the containing tanks may be pretty massive. Other less ideal substances may offer more manageable engineering problems. But the point is that while a nuclear reactor for a submarine, for example, has the enormous advantage of carrying a lot of energy in a small mass of uranium fuel, a nuclear rocket must also carry a very large mass of propellant— and much of the apparent advantage is lost. Nuclear rockets *will* be needed some day in launching very large space vehicles, and research on such rockets should be energetically pushed. But nuclear power is not a simple magical answer to all problems.

Other possibilities are being investigated—ionic propulsion, photon propulsion, alpha-particle propulsion, etc. It is too early to evaluate their practical possibilities. One thing must be remembered—no propulsion scheme, no matter how exotic, can get away from the basic momentum and energy relations. If a space object is to acquire a large velocity and if it is to escape from a gravitational field, then energy is required—and indeed the energy given to the vehicle itself is very small compared with the energy which must be imparted to the high-velocity escaping propellent. Therefore, any substantial propelling device must carry both large amounts of energy and large amounts of propellent mass. The only scheme which is not subject to these requirements is use of the radiation pressure provided by sunlight. Though this pressure is extremely small, it will, over very long periods, provide appreciable momentum. A solar-pressure "sail" can cause an object in an orbit about the sun slowly to "accelerate" (i.e., circle outward into larger orbits) or "retard" (i.e., circle inward). Since the small pressure can be available for extremely long times, an orbit gradually spiraling out to very great distances from the sun becomes possible.

2. *Local Power.* An instrumented satellite must also have energy to operate its instruments, and especially must be able to transmit the accumulated information back to an earth station. All space vehicles so far, except the Vanguard satellite, have used dry cells as local power sources—and these have become exhausted in a few days or weeks—often long before the satellite itself has returned to earth. An invisible satellite whose "voice" has gone dead is a pretty useless object.

The power requirements for the radio transmitter which is to radio information back to earth get rather imposing, as the distance from the earth increases. An earth satellite at a distance of 500 miles or so can be heard by special receivers when transmitting only 1/100 of a watt. But, as one goes farther out, the inverse-square law begins to take its toll. At 5,000 miles the power for the same receiver and same signal strength would have to be 100 times as much, or 1 watt; at 50,000 miles, 100 watts; and at the moon, 240,000 miles, one would need about 2½ kilowatts. At the distance of Mars, some 35 million miles, the power has risen to 50,000 kilowatts—1,000 times the radiated power of a normal cleared-channel broadcasting station, and approaching the power of the very largest electric generating stations now operating on earth.

A part of this difficulty can be overcome by using a directional antenna on the satellite—with the obviously difficult problem of keeping it pointed toward the earth—and by using very large receiving antennas on the earth, 100 or more feet in diameter. But, even at best, the communication problem is one of extraordinary difficulty, and even in the simplest cases one needs some sort of power supply for a long time.

What possibilities are there?

Possibly there can be better batteries. Present zinc silver batteries provide 20 watt-hours per pound of weight. It would take 440 pounds to operate a radio set consuming 1 watt continuously for one year. Using intermittent operation—transmitting only on signal from the earth—correspondingly longer times can be obtained. Theoretically it should be possible to improve this performance by a factor of about 10. For orbits near the earth, where one or a few watts of power will be sufficient, batteries will clearly be very useful. For distant ventures, however, the radio power required is so great that batteries become hopelessly inadequate and they cannot power a satellite for times of many years.

Solar power at once suggests itself—and has indeed proved its potentialities in the Vanguard satellite whose solar-powered radio was still operating a year after launching. The power level was very small however. Larger power requires larger area with corresponding engineering problems. The power available from the sun is, near the earth, of the order of 100 watts output per square meter of effective surface area for cells of present types. Thus large arrays of cells, or else large concave reflectors to concentrate the energy, will be required. For satellites which are near the earth, and hence in its shadow about half the time, some storage battery may be required—and the weight and life requirements become immediately more difficult.

An ingenious but extremely expensive device has recently been constructed in which an intense radioactive source, activated in a nuclear reactor, is used as a source of heat to activate a thermoelectric couple. Available devices might provide a few watts of power, but the most suitable isotope (polonium) has a half life of only 138 days—which cannot be prolonged even if the power is turned on only intermittently.

It would appear that unless a new invention is made, the outlook for having sizable energy sources which will supply considerable power for long periods of time and for distant journeys is gloomy indeed. So we may expect to spend millions of dollars to launch a satellite, only to have its voice fail after only a few weeks of operation—and to have it quickly fade at large distances. Here is a real problem worthy of the best developmental efforts.

It is clear that power for operation of a radio transmitter is only one requirement for an instrumented satellite. The instruments themselves, the navigation and control equipment, the cameras, Geiger counters and other equipment for scientific observation all require energy also. For these, however, the energy requirements remain constant as the object recedes from the earth, while the radio power requirement increases rapidly. In the end the radio transmitter will provide the chief problem.

If the satellite carried human beings, additional requirements arise. Food, oxygen and water will add up to substantial loads for long journeys. For orbits closer to the sun, or farther from the sun, than the earth's orbit the temperature control problem will become serious—requiring additional energy.

In summary, it can be said that space exploration opens up fantastic new vistas for research and exploration. New knowledge of the earth, of space and of our neighboring planets, hidden from human beings since the beginning of time, will soon be available. The new knowledge will be a resource of unimaginable and of unpredictable value. It will, however, be acquired at very great cost. Space is large; travel times are immense; the energy requirements for some ventures may be colossal; the technological problems will constitute a challenge to man's ingenuity for generations to come. But new inventions, designed to aid space travel, will also aid many more earthy ventures and yield new dividends to technology. These combined with the new knowledge of unforeseeable uses will certainly make space research—like all other scientific research—an exceedingly good investment.

ALAN L. DEAN

Mounting a National Space Program

DR. DU BRIDGE has provided some of the fundamental scientific facts which the United States and its agencies of government must take into account in any well-conceived attempt to explore or to wrest new knowledge from outer space. He has given us a sense of the opportunities for achievement and learning, and at the same time he has noted some of the formidable obstacles to space exploration. I am sure that it is sobering to us all to learn that if Pincus No. 7 is a planet revolving around

---

ALAN L. DEAN is the assistant administrator for management services of the newly established Federal Aviation Agency. Prior to accepting this position in January 1959, he was a senior management analyst for the Bureau of the Budget with immediate responsibility for improving government organization in the fields of science, aviation, natural resources, public works, housing, and national capital planning. In 1958, he chaired the staff group which drafted the National Aeronautics and Space Act and planned the establishment of the National Aeronautics and Space Administration. Prior to joining the Bureau of the Budget in 1947 he was director of the War Department School of Personnel Administration. He has been active in civic and community affairs, having been elected to the Arlington County, Virginia, Board of Supervisors, in 1951. He is now a member of the Arlington County Planning Commission and the Commission on Incorporation. He was born in Portland, Oregon, in 1918. He has received degrees in political science and public administration from Reed College and The American University.

Alpha Centauri, our first mission may have to travel for 30,000 years to reach this potential good neighbor. It seems that our solar system will be about all of outer space likely to be traversed by man in our generation.

As a social scientist, and more precisely a political scientist, I can add little, if any, to the reader's knowledge of space itself. I should, therefore, like to draw from some recent work while with the Bureau of the Budget to comment on two facets of our space program to which I might make a contribution. These are the problems of organizing the national space effort and of providing the financial support for what is proving to be a very expensive undertaking.

While with the Bureau of the Budget I was with the Office of Management and Organization, in immediate charge of the Bureau's work in science and aviation organization, and thus had little to do with making funds available. I have always believed, however, that our nation must take the lead in the race to understand and apply the knowledge and resources of outer space. But I felt then, as I do now, that careful planning should precede the commitment of funds and that a large space budget is not alone sufficient to guarantee American leadership in this field. Superior achievement in the conquest of space depends both on the allocation of adequate resources to the task and the establishment of an administrative and organizational framework capable of putting those resources to work swiftly and efficiently.

It may be of interest to trace how the United States has undertaken to organize for its space activities since October 4, 1957, when the U.S.S.R. placed in orbit the first man-made earth satellite. The chronicle of our achievement in the last eighteen months has been an impressive one. It is, in fact, a convincing demonstration of the nation's ability to act swiftly when the occasion demands.

The launching of the first Sputnik raised immediate questions as to why this country, with its immense wealth, its advanced technology, and its superior scientific competence, should have permitted the Russians to be first with an achievement which rivals in potential importance the voyages of Columbus or the development of the first atomic device. It certainly was not a lack of knowledge or capacity on the part of American science or industry which produced our embarrassment. It was fully appreciated before Sputnik that an earth satellite was possible with the tools at hand. The Vanguard project, which was being carried out by the Naval Research Laboratory as a part of United States participation in the International Geophysical Year, had, as one of its principal objectives, the launching of instrumented satellites in the interest of scientific progress. The National Advisory Committee for Aeronautics had been conducting advanced studies in missile research and was, in co-operation with the Department of Defense, developing the X-15, a manned vehicle designed to navigate for short periods beyond the earth's atmosphere. The military services had developed missile boosters known to be capable of projecting satellites into orbit. Yet, with all this effort and capacity in our country, the Russians were able to secure the scientific and psychological triumph of pioneering the first successful earth satellite.

Our being surpassed by the Russians can, I believe, be ascribed chiefly to the fact that we had never felt it urgent or necessary to launch a unified, high-priority space program. This deficiency was both a cause and result of the lack of a federal agency having as a primary mission the exploration of space and the advancement of the space sciences. The National Advisory Committee for Aeronautics, with all its competence, was chiefly concerned with aeronautical research and it had neither the charter nor the funds to undertake a major space program. The X-15 project was more an experiment in high-speed flight than an effort to achieve a sustained or deep penetration into space. The Naval Research Laboratory was concerned with applications of science to many facets of national defense. The Vanguard project was, therefore, but one of the pressing demands upon the Laboratory, and the funds allocated to that enterprise were relatively modest. The Defense Department's missile program was understandably focused upon devising weapons capable of fulfilling urgent defense requirements, and earth satellites did not at the time seem to add very much to the immediate promotion of national security.

Once the implications of Sputnik were assessed, the President moved swiftly to strengthen our organization for planning and executing a space program. In November 1957, he provided a focal point for planning and direction at the highest level within the government by appointing James R. Killian, Jr., President of the Massachusetts Institute of Technology, to the post of Special Assistant for Science and Technology. Dr. Killian assumed the chairmanship of the President's Science Advisory Committee, and began the formulation of a space program for the United States. The Department of Defense also strengthened its facilities for co-ordinated effort in the space sciences through the creation on February 3, 1958, of the Advanced Research Projects Agency under Roy Johnson.

The National Advisory Committee for Aeronautics concurrently undertook the development of a national research program for space technology. On January 27, 1958, it released its proposals, including the recommendation that the leadership in space science and exploration be made the responsibility of a national civilian agency. Few were surprised to find that the report envisaged the NACA as the logical agency to administer the expanded program.

The Congress also turned its attention to national needs in the fields of space and astronautics. The Senate on February 10, 1958, established a Special Committee on Space and Astronautics under the chairmanship of the majority leader, Senator Lyndon Johnson. The House followed suit with the creation of a Select Committee on Astronautics and Space Exploration on March 5. The majority leader, Rep. John W. McCormack, became chairman, and the minority leader, Rep. Joseph Martin, was included in its membership. Simultaneously, bills proposing a variety of approaches to the space problem were introduced by individual members of the Congress. The existence of the special committees and the high degree of Congressional interest in this field later proved of material assistance in making possible the prompt strengthening of the nation's space program.

In early 1958 the Bureau of the Budget, working in close co-operation with Dr. Killian and the President's Advisory Committee on Government Organization, began a systematic canvass of the alternative ways of organizing an effective space program. Its work took into account the finding of the President's Science Advisory Committee that the bulk of the space program would be civil in character. The Bureau first considered lodging the new space program in an existing agency. The Department of Defense was found to have formidable capability, but was ruled out because of the largely civil content of the anticipated program and the need to emphasize peaceful purposes. Several civilian agencies such as the Atomic Energy Commission and the National Science Foundation were also considered and not chosen for various reasons. The creation of an entirely new department or agency was found to be infeasible because of the time required to recruit and organize the staff of scientists required to do the job. The solution finally recommended was the placing of primary responsibility in a renamed, reorganized, and vastly strengthened civilian agency with the NACA as its nucleus. Dr. Killian and the Director of the Bureau of the Budget joined in making this recommendation to the President, who gave it his unqualified approval.

An Administration draft of implementing legislation was sent to the Congress on April 2, 1958, and its purposes were explained by the President in a special message. The draft bill proposed the creation of a National Aeronautics and Space Agency which would absorb the entire NACA, plus such activities of the Department of Defense and other federal agencies as the President, by executive order, chose to transfer to the new agency. The direction of the agency was to be placed under a single administrator, assisted by an advisory board. The Congress gave prompt and thorough consideration to the President's proposals, and the National Aeronautics and Space Act was approved on July 29, 1958, less than four months after it was introduced. As enacted, the legislation provided for a

National Aeronautics and Space Administration, but it also created a National Aeronautics and Space Council, chaired by the President, and containing as members the Secretary of State, Secretary of Defense, Administrator of the NASA, Chairman of the Atomic Energy Commission, Director of the National Science Foundation, and three citizens from private life.

The Council was intended primarily to help solve a problem which arose in connection with the preparation and Congressional consideration of the space legislation; namely, that of distinguishing the responsibilities of the new NASA from those appropriate for retention in the Department of Defense.

It is self-evident that the Defense Department, with its missile programs and its concern for the security of the United States, has a valid basis for participation in research and development in the space sciences. On the other hand, the President had stressed the importance of a civilian setting for the space program of the United States. The law, as approved, provided that the government's aeronautical and space activities should be the responsibility of the civilian agency except that "activities peculiar to or primarily associated with the development of weapons systems, military operations, or the defense of the United States (including the research and development necessary to make effective provision for the defense of the United States) shall be the responsibility of and shall be directed by the Department of Defense." It was anticipated that disagreements would from time to time arise as to what space activities were peculiar to, or primarily associated with, defense. The legislation therefore empowered the President, with the assistance of the National Aeronautics and Space Council, to determine which agency should have responsibility for any space activity.

It will surprise no one familiar with the problems of government administration to learn that difficulties have indeed arisen between the Department of Defense and the NASA. A few months ago the nation's newspapers carried detailed reports of the dispute over the proposed transfer of jurisdiction over the Jet Propulsion Laboratory and the Army Ballistics Missile

Agency from the Department of the Army to the new Administration. After much controversy and negotiation, the President decided that the Jet Propulsion Laboratory should be placed under the supervision of the NASA. However, the Department of the Army was permitted to retain primary control over the Army Ballistics Missile Agency. There has also been uncertainty as to whether a number of space programs still being carried out under the direction of the Advanced Research Projects Agency should not be placed under the direction of the civilian agency. There remains some danger that confusion as to the roles of the NASA and the Department of Defense will delay or make more costly important space projects.

There is every reason to hope, however, that accommodations will be reached between the Department of Defense and the NASA which will assure a mutual strengthening of each other's work rather than a destructive rivalry. It should not be forgotten that the predecessor of the NASA (the NACA) performed most of its work in areas having direct military applications and did so in such a manner that its findings were immediately available for use in weapons systems and other military projects. The NASA, in addition to discharging its responsibility for the civil space program, should be able to use its laboratories and its research contracts to meet some of the special needs of the Department of Defense without the duplication of costly facilities.

A completely satisfactory organization for the space program of the United States does not exist today, and it may never be achieved. We have, however, in a very short time created a framework within which an adequate space effort can be planned and executed with due regard to both civil and military requirements.

Improved organization greatly facilitates technical and financial planning, but it does not of itself resolve a question which will be with us for many years, namely: How much of the total resources of the nation should we devote to the conquest of space and the support of contributing scientific research? We all know that the federal budget represents the taking of wealth from our citizens to finance public purposes. We now have a gross national product in the order of $440 billion. The federal government is expected to spend some $81 billion during the 1959 fiscal year, an amount in excess of 18 per cent of the gross national product.

It is now estimated that the total federal expenditures for research and development in the fiscal year 1959 will approximate $4,841 million. In 1960 this amount is expected to increase to $5,484 million. About one-tenth of the total is for space activities. In 1959, the NASA is expected to spend $153 million in carrying out its programs, and in 1960, its total expenditures will rise to approximately $280 million. The Department of Defense expenditures for space activities are estimated at approximately $174 million for 1959 and $305 million in 1960. Certain other agencies such as the Atomic Energy Commission are investing smaller amounts in projects having potential space applications. The sums now being devoted to the government's space programs are substantial, but are still modest when compared to the total resources available to the nation. However, expenditures tend to be small in the initial stages of programs involving extensive research, and to increase in later years. Dr. Glennan, the NASA administrator, not long ago estimated that annual cost of the space program will rise to at least $1 billion annually. That his prediction may be a conservative one is illustrated in that total new appropriations for space research requested in the 1960 fiscal year budget are estimated at $881 million.

At present the conquest of space has a strong appeal to the public and vigorous support from the Congress. We hear frequent demands for a more rapid expansion of our space programs, and such demands are often supported by persuasive arguments. In weighing the merits of appeals for more money for space we should not, however, forget that even in the field of scientific research there are numerous other ways to use our limited tax resources. For instance, diseases that kill or cripple millions of our citizens annually might be overcome by more elaborate and costly research programs. It can, therefore, be argued that the government should spend its funds on the control of heart disease, arthritis, cancer, and mental illness, instead of seeking to learn what lies under the cloud cover of the planet Venus. We also know that in such fields as oceanography the Russians have been pushing ahead vigorously and have equipment and programs which may be considerably in advance of ours. Agricultural research to control plant and livestock diseases and to speed development of more productive strains of plant and animal life assumes an increasingly urgent aspect as the world's population grows. I mention these opportunities only to stress that the Bureau of the Budget and the President, in considering requests for funds for space exploration and research, must take into account equally attractive competing uses of necessarily limited public funds. If the advocates of ambitious space programs occasionally encounter difficulty in obtaining all the money they desire as quickly as they would like to have it, this does not necessarily mean that the Bureau of the Budget and the Administration are unappreciative of the potential contribution of the space sciences to the security, prestige, and well-being of our nation. It is simply that officials charged with financing a myriad of essential programs must strive for a balance which recognizes true priorities. A responsible approach to government leaves no other choice.

PHILIP C. JESSUP

The International Opportunity

PRESIDENT DU BRIDGE, in presenting the scientific and technological aspects of our problem of utilizing the resources of outer space, stresses his belief that the greatest resource to be obtained is knowledge. He warns that the planning of programs must be done by scientists and engineers and not by persons "whose only source of information is the comic strip." I am not scientist nor engineer nor politician, although I may share the lack of expert knowledge of the last category. Yet I venture to

PHILIP C. JESSUP, Hamilton Fish professor of international law and diplomacy at Columbia University since 1946, and co-chairman with Professor Rabi of the University's Council for Atomic Age Studies, has had a distinguished career in both the academic and governmental fields. He began to teach international law at Columbia in 1925 and became professor in 1935. He has served as legal adviser to federal government offices in numerous instances, at international conferences, embassies, and to United States delegations at the United Nations. He represented the United States in the United Nations Security Council and General Assembly, in various sessions from 1948 to 1952. He was appointed Ambassador-at-Large in 1949, and resigned in 1953. He is the author of several books, including a two-volume biography of Elihu Root (1937) and *A Modern Law of Nations* (1947). He served as the supervising editor of Columbia University Studies in History, Economics, and Public Law, 1929-33; and is a member of the board of editors of the *American Journal of International Law* and of *International Organization*. Mr.

ask the question which was the title of a book by my colleague Robert S. Lynd some years ago—"Knowledge for What?"

I would suggest that the knowledge we need from this great enterprise of our exploration of outer space is knowledge for peace rather than knowledge for war, but that, actually, the endeavor to obtain knowledge for war is at the present time much more highly organized than is the search for the utilization of space knowledge for peace. We know relatively little about managing man's stupidities here on earth but perhaps, with the advantage of science and technology, we may learn more rapidly about the utilization of outer space for peaceful purposes.

Now, a small and inadequate sampling suggests to my inexpert mind that even scientists and engineers are human and, therefore, I suppose they disagree among themselves; and when one talks about what one may anticipate in terms of developments in outer space—accepting 110 per cent what Dr. DuBridge has said about the scientific and technical possibilities and difficulties—I do suppose that the rate of progress with these various space developments will be a function of the size of the effort; that is, of the budget and the extent of the mobilization of human resources, and the priorities in general. I suppose, also, that those priorities will be determined largely by military considerations; and military considerations are not considerations of economy. They may not even be considerations of wise scientific planning. I'm not much concerned with any exact timetable. I think we can assume from various official statements that certain events will take place in x years; it is not essential for the purpose of this discussion to determine now whether x equals 5, 10, or 20 years, or 20 plus y. At the same time, a layman in matters scientific must pay some heed to

Jessup was born in New York, N. Y., in 1897, and received his A.B. degree at Hamilton College, his A.M. and Ph.D. degrees at Columbia University, and his LL.B. at Yale. He is a member of the bar of the District of Columbia and of New York, and was for sixteen years a partner in a New York law firm.

official statements of pending programs. For convenience, take a document recently published by the House Select Committee on Astronautics and Space Exploration—a document called *Space Handbook*—prepared by the staff of The RAND Corporation. The anticipations expressed in that book are very much in line with recent official estimates prepared by the National Aeronautics and Space Administration, which we are learning to call NASA.

I repeat these anticipations without any personal judgment or appraisal because I am not competent to have any, but according to the estimates in the *Space Handbook,* this is what we may expect within the next five years in terms of anticipated government effort:

Orbit satellite payloads of 10,000 pounds at 300 miles altitude.
Orbit satellite payloads of 2,500 pounds at 22,000 miles altitude.
Impact 3,000 pounds on the Moon.
Land, intact, 1,000 pounds of instruments on the Moon.
Land, intact, more than 1,000 pounds of instruments on Venus or Mars.
Probe the atmosphere of Jupiter with 1,000 pounds of instruments.
Place a man, or men, in a satellite orbit around the Earth and recover after a few days of flight.

Now, according to the first Annual Report of NASA, which was issued in February 1959, it is planned to have "fixed" satellites in space by 1960 or 1961. These are not, as I understand it, designed to float lazily as Dr. DuBridge says they couldn't do. Instead, NASA says: "These devices will maintain a fixed position over a given point, revolving at the same speed as terrestrial rotation, at a distance of 26,000 miles from the center of the Earth. Three of these satellites could relay radio, television and teletype messages continuously." The Advanced Research Projects Agency has announced that these satellites will weigh many tons. I do not know what their expected life or lives will be. You will recall that President Eisenhower on March 3 pointed to some of the emerging problems in a message asking Congress for authority to appoint a telecommunication commission which, among other things, would restudy on the national level the allocation of radio frequencies to avoid conflict between government and nongovernmental use, involving also contact with satellites. If this is done, it will facilitate the solution of some of the international problems to which we may look forward.

To go back to the estimates of this year, 1959, NASA plans navigation satellites which will supply a precise all-weather system for determining sea or air positions anywhere on the globe to be used by aircraft, surface vessels, and submarines. The first ones will be, as Dr. DuBridge suggests, in many instances short-lived and therefore will not have permanent value, I presume, for their navigational mission. Orbiting weather satellites are expected by the meteorologists to supply data which will aid them in weather prediction—the first ones having already been launched, again with short lives. I don't find in any of these official statements that they have gone too wild on the comic strips. I don't see anything about planting whole colonies of people on Mars or Venus or some other planet, or even the popular notion that we're now going to be able to control the weather. And I haven't seen any project for organizing a coal mining company to work on the Moon or Mars. I wouldn't be sure we wouldn't get it in the presidential campaign, though.

Whether or not it makes sense from the point of view of science or of politics or of military preparations, competition with the Russians will, in my opinion, continue to be a prime factor in our space program. It is officially stated by the United States Government that the Russians have the capacity to put man in space before we shall be ready to do so, but our man-in-space program, which is known as Project MERCURY, has the highest priority among all our civilian space projects. Here we will enter, as Dr. DuBridge says, a new phase which stimulates the imagination, conjures up all sorts of impossible things, and yet, I believe, offers a perfectly extraordinary challenge to mankind.

Now if we still follow as our guide the *Space Handbook*, it is indicated that on the military side satellites with nuclear warheads could be launched without our having knowledge of the launching, and maintained on distant orbits long in advance of war until they were recalled and directed to a specific target on Earth. I don't know whether the timetable is 1, 5, 10, or 20 years. More immediate perhaps, and considered very seriously from a military point of view, is the possibility of interference with military attack-warning systems by jamming through satellites equipped for that purpose. The possibility of countermeasures to jam the jammer is also envisaged, and then, affirmatively and hopefully in a rather gloomy military picture, the *Space Handbook* points out that "opportunities for international inspection through astronautics could be of incalculable value to world peace and security." They suggest, and support reasonably well as far as I can tell as a layman, that if many countries can develop this capacity (the Soviet Union and the United States will surely not long have a monopoly in rockets, missiles, and satellites) it may affect the trend of discussions, such as those now going on in Geneva, on the method of inspection to detect nuclear tests, and might, if they come in time, or if they are anticipated with enough certainty, equally influence a resumption of the talks now broken off concerning the regulation or the guarding against surprise attack.

Now, those considerations lead us to consider what international measures could usefully be taken to deal with outer space or with man's increasing activities in outer space. Last December (1958) the General Assembly of the United Nations adopted a resolution which the United States had joined with nineteen other countries in sponsoring. This resolution, which was finally passed, provided for establishing a committee on peaceful uses of outer space. As so often happens in these matters, we got into a wrangle with the Soviet delegation about the composition of the committee. Not being able to reach agreement with them on what countries should be on the committee, the Soviets and the Czechs and the Poles, who were all elected to the committee, have announced they won't participate. Since this was designed to be a measure to bring about some co-ordinated results, the committee is not going to serve the purpose originally contemplated. Nevertheless, it has been created and is charged with reporting to the General Assembly next September, and these are the things it is supposed to cover: first, "The activities and resources of the United Nations, of its specialized agencies, and of other international bodies relating to the peaceful uses of outer space"; second, "The area of international co-operation and programmes in the peaceful uses of outer space which could appropriately be undertaken under United Nations auspices to the benefit of states irrespective of the state of their economic or scientific development . . ."; third, "The future United Nations organizational arrangements to facilitate international co-operation in this field . . ."; and lastly, "The nature of legal problems which may arise . . ."

Now I think the question one must ask oneself is whether all of these problems could be more efficiently handled by a continuation of the type of co-operation established in the International Geophysical Year—the well-known IGY. That very successful effort was launched by the International Council of Scientific Unions, and it is worth noting that while the General Assembly of the United Nations was haggling over the composition of the committee it was setting up, this same International Council of Scientific Unions was holding a meeting with Soviet participation here in Washington, and with no difficulty at all established a new Committee on Space Research, known as COSPAR, which is going forward on an international scientific co-operative basis following the general pattern of the co-operation of the IGY. (It should be noted, though, that problems of political representation also soon emerged in COSPAR.) Undoubtedly there are some things that can be done better through an intergovernmental co-operation. The Legal Adviser of the State Department has said that in connection with the peaceful uses of outer space it is "axiomatic

that some appropriate international machinery be created."
These aren't very elaborate or magical things. They are practical things. For example: one of the most immediate is international agreement on the allocation of radio frequencies for communication with satellites. On an earth basis this has long been done by the International Telecommunications Union, one of the specialized agencies of the United Nations, not always without friction but with considerable success. The ITU could be used to deal with similar satellite problems.

We have also the problem of the choice of orbits and times of launchings. That is an important problem, the solution of which would seem to have a mutual interest for all countries concerned. It is not fantastic to foresee a crowding of the spaceways. There is certainly a very considerable economic wastage at the present time in not having any agreement on what satellites are going to be launched with what instrumentations for the accomplishment of what particular missions. If co-operation could be established it certainly would be helpful and would be more economical. We do not have any international organization at present which has had that kind of function, and probably some new international organization would be needed for that purpose. The same organization probably could be designed for co-operation in the business of tracking satellites. We had difficulties in tracking Pioneer IV. The Russians had similar difficulty in being unable to track their "moon shot." A system of internationally co-operating tracking stations would be of value in all enterprises of that kind.

So, too, as weather satellites are put in orbit, if we could establish tracking and computing stations widely spaced for both polar and equatorially orbiting satellites, a great deal could be done to facilitate the translation of information for use by the meteorologists in weather predictions which would be of advantage to various countries, in fact, to all countries and to aerial navigation and otherwise all over the face of the earth. But this, too, requires international co-operation, and countries which are quite incapable themselves of getting into space activities in the sense of launching satellites—like a small country in equatorial Africa—could be of great value in permitting the establishment on its territory and facilitating the operation of a tracking and computing station in that location which might be advantageous. In these matters, another United Nations specialized agency, the World Meteorological Organization (WMO), could be helpful.

Can you do these things internationally? One is tempted to recall that for ninety years (until just a couple of years ago when a new system was installed) the Cape Spartel Light, which lights the southern entrance to the Mediterranean Sea, was operated under a treaty by international agreement and on an international scale.

Now I have mentioned that the United Nations Resolution also calls for an examination of the legal problems. As a lawyer it seems to me that a good deal of the voluminous legal literature on this subject is not helpful. There is a great deal of discussion about how high is the sky, and where does the air stop, because we have treaties that refer to air space and they are not applicable to the outer reaches beyond the air space. Some day we may have to define them. But I don't think that is the primary job to be done. I do believe that it would be extremely useful, politically and legally, if we could secure agreement at this time, as the Secretary General of the United Nations has urged, that all of outer space is the common property of mankind; that no national claims can be established to outer space or to the planets. But government officials feel the responsibility to guard their claims on the chance that some new developments will give to their state a military advantage which should not be signed away in advance. Thus we find the Legal Adviser of the Department of State asserting that the inherent right of self-defense, which is recognized in Article 51 of the United Nations Charter, is equally applicable in outer space. One cannot by any means exclude the possibility that some state will lay claim to the Moon as a result of a first hard or soft landing on the Moon's surface. Whether international agreement is reached or not, it is, I think, quite obvious that these problems are not going to be solved by the application of earth-developed and earth-oriented legal principles or legal theories. They are going to be worked out on the basis of new international agreements, compromises in which eventually we must have a common basis, a common approach with the principal other country concerned—the Soviet Union. It is only through this kind of co-operation which can be achieved through the United Nations that this knowledge of outer space can be successfully utilized for peaceful adaptation.

Index

Acetylene: hydrocarbons 117, 125; petrochemicals 117; production overseas 140

Ackerman, Edward A. xiii; biographical note 63fn-64fn

Adams, John A. S.: biographical note 75fn; minerals — exploration eliminated? 95-96, shortages, exhaustion? 93; uranium industry 98

Adenine: DNA structure 12-13

Adhesives: waterproof, synthetic 119

Advanced Research Projects Agency 222, 225, 230

Advertising: chemical firms 132; chemical research, balance 118

Advisory Committee on Weather Control: cloud seeding 46; rainmaking 58; water-use planning in west 64

Africa: food protein syntheses 143; uranium 156

Agriculture: chemical industry 116; efficiency *vs.* city 23; new multipliers (title) 29-34; research 237

Air pollution: cloud modification 65

Aircraft: navigation satellites 231

Alaska: natural resources 142-143

Alcohols: synthetic 140

Algae: Japan 143

Aluminum: alloys 126; clays, extraction 89; manufacturing outside U. S. 140; organometallic compound 126; plastics 121; price increases 109; raw materials 102

America: discovery, population growth 23

American Association for the Advancement of Science 67

Amino acids, *see* Phenylalanine.

Anderson, Clinton P.: biographical note 54fn

Anemia: sickle cell 7

Animals: hybrid 110; Mendelism 5

Antennas (Radio): satellite 216

Anthracite coal: supply curve 103

Anti-freeze solutions: ethylene glycol 124

Antibiotics: production overseas 140

Arco, Idaho: National Reactor Testing Station 158

Arctic Ocean: ice pack disappearance 52-53

Argentina: petroleum industry 98, 141

Argon: atmosphere free source 88; exploration reduced 90

Argonne National Laboratory 189-190, 191, 193

Argus experiment (Mar. 26, 1959) 207fn

Arizona, University: cloud formation studies 50

Army Ballistics Missile Agency: 224-225

Arsenic organic compounds: valuable drugs 126

Asia: food protein syntheses 143; Russian industries 143

Atmosphere: atomic detonations 59; earth, radiation 206; global, warming 52; hydrogen weapons fallout 59; national scientific program 68

237

Milton Keynes UK
Ingram Content Group UK Ltd.
UKHW050304161024
449569UK00033B/311